HUMANS, ANIMALS, AND THE CRAFT OF SLAUGHTER IN ARCHAEO-HISTORIC SOCIETIES

In this book, Krish Seetah uses butchery as a point of departure for exploring the changing historical relationships between animal utility, symbolism, and meat consumption. Seetah brings together several bodies of literature – on meat, cut marks, craftspeople, and the role of craft in production – that have heretofore been considered in isolation from one another. Focusing on the activity inherent in butchery, he describes the history of knowledge that typifies the craft. He also provides anthropological and archaeological case studies that showcase examples of butchery practices in varied contexts that are seldom identified with zooarchaeological research. Situating the relationship between practice, practitioner, material, and commodity, this imaginative study offers new insights into food production, consumption, and the craft of cuisine.

Krish Seetah is an assistant professor of anthropology at Stanford University. He brings vocational experience as a professional butcher to his research on environmental archaeology, especially human–animal interactions, and the ecological consequences of colonialism. He is the co-editor of *Bones for Tools – Tools for Bones: The Interplay between Objects and Objectives* (2012) and editor of *Connecting Continents: Archaeology and History in the Indian Ocean* (2018).

HUMANS, ANIMALS, AND THE CRAFT OF SLAUGHTER IN ARCHAEO-HISTORIC SOCIETIES

KRISH SEETAH

Stanford University

CAMBRIDGE
UNIVERSITY PRESS

University Printing House, Cambridge CB2 8BS, United Kingdom

One Liberty Plaza, 20th Floor, New York, NY 10006, USA

477 Williamstown Road, Port Melbourne, VIC 3207, Australia

314–321, 3rd Floor, Plot 3, Splendor Forum, Jasola District Centre, New Delhi – 110025, India

79 Anson Road, #06–04/06, Singapore 079906

Cambridge University Press is part of the University of Cambridge.

It furthers the University's mission by disseminating knowledge in the pursuit of education, learning, and research at the highest international levels of excellence.

www.cambridge.org
Information on this title: www.cambridge.org/9781108428804
DOI: 10.1017/9781108553544

First published 2019

Printed in the United Kingdom by TJ International Ltd. Padlow, Cornwall

A catalogue record for this publication is available from the British Library

Library of Congress Cataloging-in-Publication Data
NAMES: Seetah, Krish, author.
TITLE: Humans, animals, and the craft of slaughter in archaeo-historic societies / Krish Seetah, Stanford University.
DESCRIPTION: Cambridge, United Kingdom ; New York, NY : Cambridge University Press, 2019. | Includes bibliographical references and index.
IDENTIFIERS: LCCN 2018023526| ISBN 9781108428804 (hardback : alk. paper) | ISBN 9781108447317 (paperback : alk. paper)
SUBJECTS: LCSH: Slaughtering and slaughter-houses–History. | Meat industry and trade–History.
CLASSIFICATION: LCC TS1960 .S44 2019 | DDC 338.4/76649029–dc23
LC record available at https://lccn.loc.gov/2018023526

ISBN 978-1-108-42880-4 Hardback

CONTENTS

FIGURES

TABLES

PREFACE

FROM APPRENTICESHIP TO SCHOLARSHIP

The naissance of this book began well before I had completed my GSCE exams (British national examinations undertaken as part of secondary school education, from age fourteen to fifteen years). One month after my thirteenth birthday I was offered a job in a butcher's shop, in Brixton, South London. My qualification for the job was that I was 'tall for my age' (which I wasn't). I worked an average of twenty-five to thirty hours per week throughout my secondary and sixth form college education, and the initial period of my degree in biology, until the age of twenty: seven years, the length of time a butcher's apprenticeship has lasted since the medieval period. The knowledge and experience I gained during this period were crucial guides to the theses I undertook for my master's and doctoral studies. I would like to say that I planned my academic career around this past knowledge, having realized that a deep working knowledge of modern butchery would be instructive for my archaeological studies. This could not be further from the truth. The serendipity of meeting and subsequently working with Mark Maltby at Bournemouth will remain, for me, evidence of Fate. At the point of entering into a master's degree at Bournemouth, I initially commenced on a course in forensic archaeology and biological anthropology. However, my interests were focused as much on faunal as on human morphology, which was in fact the reason why, as someone growing up in urban London, I had accepted the job as a butcher's apprentice. To my young mind, this seemed a way to observe animals, albeit dead ones. Thus, I approached Maltby, as program convener of Osteoarchaeology, to ask if I might join the course. The circumstances of this interview were troubling. With the exception of my biology courses, I did not have any background or knowledge of archaeology, let alone osteoarchaeology (nor did I have the remotest idea of Maltby's own work). In desperation I mentioned that I had been a butcher and knew about animal anatomy from that perspective. The remainder of the narrative is self-evident. Had I known that Maltby had been working on Romano-British *butchery* since at least 1979, I might have led with that argument!

I explain these details in order for the reader to understand three things. The first is that while the knowledge of modern butchery might seem singular, it is only part of what I brought to the study of archaeological cut marks – and what is needed. Biology, in the guise of animal physiology, formed a major part of my school and undergraduate education. I was learning about muscle groups, attachments, and morphology in tandem with physically working with animal bodies in a professional capacity as a butcher. I also undertook a master's course in Ecology for Sustainable Development, with numerous modules focused on indigenous and folk knowledge and taxonomy. This training provided a usable point of departure for teasing out the details of human-environmental relationships and has been highly beneficial for the ethnographic studies I have been involved in. Thus, this academic diversity has provided an operational knowledge base that I have tapped into consistently throughout my archaeological research.

Second, the above should illustrate the specific alignment of circumstances that brought me to archaeology. In so doing, it should also demonstrate how and why it is relatively rare that those with vocational knowledge make the transition to academia. Modern-day meat trade butchers are generally not interested in contributing to archaeological debates. Furthermore, despite having a well-developed working knowledge and utilising animal morphology, they do not approach the subject from the perspective of animal biology. In short, people from different walks of life enter these two paradigms of learning, vocational versus academic. This is not to imply that my case is unique. Derrick Rixson put the many years of professional engagement as a butcher to good service in both his publications on meat trading and butchery (1988, 2000). Nor is this example specific to archaeology. Dominic Pacyga drew heavily from his own personal experiences of working in the meat trade, and that of his family, in his book *Slaughterhouse* (2015), an insightful historic account of Chicago's meat packing industry.

Finally, the above is offered in part to help explain why it has taken quite some time for this book to see the light of day. During my PhD, I spent relatively little time thinking about the meaning of butchery; instead, I recorded cut marks and drew conclusions. Despite consistently reading and being informed that butchery was an important aspect of zooarchaeological research, that there were inadequacies in the methods used, that the subject (of which I should have cared about deeply) was receiving less and less attention, I was still not motivated to *think* about butchery. This obtuse mindset stemmed from the time I worked in a professional capacity. Although working as a butcher, I had no aspirations to stay in the meat trade. My ambition had always been to pursue a career in academia. Thus, butchery became a means to an end, not an end in and of itself. Ultimately, it served as an idiosyncratic path into an academic career. It has taken time, and much astute mentorship from colleagues near and far, to realize my error. For all my apparent expertise in the practice, I had very little sense of what zooarchaeology required, and what I might bring to the study of cut marks. Indeed, I had a tendency to disregard my own past knowledge and tried to think in terms of an academic understanding of practice. The situation was a quintessential example of the blind leading the blind. I couldn't see the conceptual gap between my knowledge and the current use of cut mark recording to interpret butchering behaviour. I began to understand the extent of this gap from a range of sources, and often indirectly. One revealing situation followed an article review. The referee pointed out my naivety at an assumption I had made in the manuscript, suggesting that I had failed to realize that analysts were recording what they saw, not the behaviour. Eureka! This statement revealed with alacrity the fundamental error in my view of the mechanics of cut mark recording. I did indeed believe faunal analysts were recording actions and behaviour – and in fact, I was. However, this realization also revealed an academic sleight of hand. If analysts were not recording actions and behaviour, what exactly were they recording? This served as the first domino to topple in a long chain. I began to understand some of the more fundamental issues, many of which had been long recognised but remained unresolved. A discussion of these issues forms a major component of Part II of this book.

Coming back to knowledge: one never stops being a butcher. There is a sensory legacy that cannot be erased: the smell of the butcher's shop, a coalescing of sawdust and raw meat, comes to my mind whenever I hear Mike and the Mechanics' 'The Living Years' – the song that was playing on the radio the first time I walked into the butcher's shop to start work. Over time, my olfactory sense became entirely inured to the smell and it was only ever brought back as memory when I heard the song.

Nor can the training be unlearnt. I cannot help but see animals in their constituent parts; I never think about *how* to butcher an animal, the knowledge is now institutionalised as action: psychology and physiology work in tandem without significant conscious direction. Thus, during my early research analysing cut marks it was a straightforward process to interpret archaeological butchery and deduce what had taken place in the past. I tested this model. I gave a number of bones to my former colleagues in the butchery profession; to my surprise their conclusions resonated with my own. The harder task was in explaining my observations and bridging the gap between practice and theory – which has formed the basis for Chapters 4 and 5. Recording my efforts was an even greater challenge, one that I faced since I started looking at cut marks for my master's thesis in 2002. Initially, I tried using the systems already in place. These protocols did not provide an appropriate way of recording my observations, so I developed my own form during my PhD. I amended very little from my PhD supervisor Preston Miracle's recording scheme, which I had been using extensively. However, even the small changes I made were undertaken in a mechanical way. I did not give any thought as to why I needed new parameters; I merely added them based in part on intuition and trial and error. In time, and as a consequence of working on a greater variety of assemblages, I went back to using the recording database developed by colleagues at Bournemouth University. This Access-based system allowed for subforms and served as a major departure from the recording protocols I had used previously, providing the breadth to record the dimensions of butchery that reflected on behaviour. The butchery subform, which underpins the main topics and case studies outlined in Part II of this book, permitted a simple yet important development. It allowed for an effective way to record and therefore better understand the relationship between the tool used and the outcome in the form of a cut mark, effectively leading to better interpretations of butchery practice. Other basic issues also surfaced during my PhD, particularly with regard to terminology, noting, for example, the interchangeability of 'chopper' and 'cleaver' (the latter a more accurate description of a specific tool, as opposed to an action), which could lead to ambiguity and misrepresentation, as well as subjectivity in interpretation.

Despite identifying numerous lacunae and how to resolve these gaps, it has only been through the process of writing this book that I have come to realize the true complexity of butchery and butchering – terms that I unpack and define in Chapter 2.

I had been guilty of not taking the time to think through the problem, of seeing the misconceptions that surrounded the subject matter but assuming they were specific to individuals, rather than an outcome of deeper-seated problems with our methods and approach to interpretation. Viewed from the perspective of knowledge, forming the driver for Part I of this book, we have a lens on the richness of the subject matter, placing greater emphasis on the behaviour associated with carcass processing rather than on the outcomes of butchery practice. This stance also initiates the process of addressing a long-held concern: that primacy has been placed on the minutiae of the cut mark and not on the wider meaning of butchering. However, focusing on butchery and butchering, and not the infinitely more salient topic of meat, has been a challenge; even my reviewers posed the questions 'I assume you're talking about meat and products at some point . . .' and 'Where in the book will you cover meat by-products?' This exemplifies the deep-held notion of butchery and butchering as equating to meat, which of course it does; however, that is not the sum of the subject. Once we observe butchery as an activity like any other, and butchering as behaviour like any other, we can appreciate the inherent limitation in concatenating cut marks, butchery, butchering, and meat. We wouldn't study glass production by looking only at beads, or metallurgy by looking only at knives. To understand the entire picture, we need to reconceptualise the broader topic, separate it from meat, and situate it within its proper context.

Without euphemism, but a large measure of humility, I hope the reader will agree that this book is needed. Butchery – for all its promise as a remarkable window into the past – is increasingly becoming a dead end. I am convinced that most zooarchaeologists would agree that there has not been a clear conceptualisation of the subject and that debate and questions around the topic have stagnated. Nor is this situation singular to archaeology. Different disciplines have focused on aspects of butchery; however, none has carved out the subject matter in its own right nor situated the knowledge. Although I discuss the products and outcomes of butchery, the book complements but is fundamentally different to the extensive literature on taphonomic investigations of butchery, economic studies using cut marks, or writings on meat. It is my aspiration that this book and the research that underpins it can emulate and follow in the footsteps of Eric Higgs. One underlying aim is to bring to archaeology, a discipline that focuses on the progression of human day-to-day life, valuable layman knowledge in a form that is accessible, academically supported, and can be used to advance the theory and practice of the subject.

Ultimately, to appreciate the complexity of butchery, more attention is needed on the many ways in which this activity fits into society. Such an endeavour cannot be undertaken from a single viewpoint. Moreover, 'butchery' is more than one topic. For these reasons it has been necessary to draw on a wealth of source material – anthropological, ethnographic, historic, and archaeological – in order to stimulate our awareness of the possibilities of butchery to inform us about past (and present) social practice. Thus, for this book, archaeology brings the technological context, and anthropology and ethnography illustrate the social role. In addition, historical sources alongside contemporary analogues from the modern industry provide details of economic production and how one learns and applies the craft today. From this multidisciplinary standpoint, we can start to tease out the nuances and contours of this remarkably rich source of data.

ACKNOWLEDGEMENTS

I have many individuals to thank but will start with institutions and funding bodies that have supported my efforts throughout the processing of gathering the data, and generating the ideas, that have made their way into this book. In 2003, I was exceedingly fortunate to be awarded a Research Scholarship from Peterhouse, Cambridge, to complete my PhD. I will always be in debt to the Master and Fellows of the college for this life-changing fellowship, but also for all that college life at Peterhouse had to offer. I hope that this book can serve as a small recompense. Creating the medieval cleaver formed part of a project carried out at the Lejre Centre, Denmark, in August 2003. The project took advantage of the Lejre Centre's excellent facilities, including its forge, and was very kindly funded by the centre. Numerous other small grants allowed me to undertake overseas travel for research purposes and to disseminate results, including awards from the Dorothy Garrod Fund, Fitzwilliam Trust Research Fund, H. M. Chadwick Fund, Worts Travelling Scholars Fund, Cambridge Philosophical Society, F. S. Salisbury Fund, Kathleen Hughes Fund, Ridgeway-Venn Fund, Anthony Wilkins Fund, Emslie Horniman Fund, Council for British Archaeology Publication Grant, Prof. Dame Elizabeth Hill Fund, Hugh Last & Donald Atkinson Fund, and the Royal Historical Society: Grant for Individual Travel. More recently, I have been extremely grateful to receive a grant from the Lang Fund for Environmental Anthropology, Department of Anthropology, Stanford, in support of ethnographic research in Kenya; to be selected for the Manuscript Review Workshop, Stanford Humanities Center, Stanford, which was an invaluable opportunity for me to receive feedback and critical commentary on the manuscript; and to be awarded a prestigious Hunt Post-Doctoral Fellowship from the Wenner-Gren Foundation, to complete this manuscript.

Given the nature of this book, special thanks are due to a group of individuals who have provided assistance, and indeed knowledge and training, with regard to practical aspects of butchery. One of the more interesting aspects of my academic research was in fact entirely applied and focused on the creation of cleavers from the Roman and medieval periods. I will forever be grateful to Hector Cole and Aaron Hvid for generously sharing their knowledge and craft skills during two separate periods spent working with them creating a range of knives. However, where practical matters are concerned, my most sincere and deepest thanks are due to my 'boss' Paul Salih, and to Andy George, Eddie Hakki, and colleagues at Atlantic Meats, still operating in Brixton, South East London, at the time of writing. I learnt the craft of butchery through these individuals, and much more. Needless to say, without their training, in particular and especially the patient guidance of Paul Salih, this book would simply not have been possible.

I am profoundly indebted to members of the Maasai community who operated the Kuku Game Reserve, Oliotokitok, Kenya, for their assistance with the ethnographies that have become such an important part of this book. James Maantoi in particular showed a willingness to discuss a number of topics that are deeply rooted in Maasai cosmology. I am also very grateful to Daniel Mabuvve, sadly now deceased, who expertly guided me through the processes that underscore Maasai butchery. Finally, I am very pleased to still be working with my close friend and colleague Jackson Killinga, who since 2010 has been of immeasurable support within the Maasai community and beyond in the wider region of Kimana.

At the meeting point of vocational and academic learning are two especially important and noteworthy colleagues, Mark Maltby and Preston Miracle, respectively, my Master's and PhD theses supervisors. They have provided the essential means and support to help me bridge the gap between practical and scholastic aspects of my research.

I can't imagine that any single authored book is ever actually the work of a single author! Certainly, in my case, this book attained balance through the comments and advice of my colleagues and reviewers. I am deeply indebted to Daniel Sayer for his insights on the nuanced role of the clergy in medieval Britain, to Ryan Rabett for his clear views of the ways in which archaeological and modern contexts could show resonance – 'hunting and gathering in supermarkets' – and to Alan Outram for feedback on earlier experimental archaeological research, which has since been useful for this book. Many individuals reviewed chapters of this book, and I am very grateful to Saša Čaval, Diego Calaon, Aleksander Pluskowski, Ryan Rabett, Mark Maltby, Richard Roberts, and Bryan Hanks for providing many insightful comments that helped improve the manuscript. Intellectually, the book was refined considerably following the Manuscript Review Workshop, and I am immeasurably grateful to Preston Miracle, again, and to Bryan Hanks, Mike Shanks, Pam Crabtree, Jonas Nordin, and Guy Geltner for the detailed and thoughtful commentary they provided during the workshop. I am also indebted to two anonymous readers solicited by Cambridge University Press for their comments on the final draft version of the text. In terms of the structure and flow of text, I will forever be equally amazed and grateful to Ian Simpson, and especially Kelda Jamison, for their capacity to draw out both meaning and nuance from the text I produced.

In addition to their academic guidance on this book or aspects thereof, I am also enormously grateful to Laszlo Bartosiewitz, Alice Choyke, and Aleksander Pluskowski for their unwavering belief in the merits of producing a book on butchery; their conviction formed the basis for my own.

Since joining Stanford, two vitally important and remarkable colleagues have provided not only much needed guidance on all aspects of producing this book, from conceptualisation of the idea to the process of bringing a book to press, but also resolute and definitive mentorship on all aspects of building one's academic career. I am humbled and honoured to be able to count Ian Hodder and Lynn Meskell as my guides and mentors.

I owe an especial and heartfelt debt of thanks to Saša Čaval, who shouldered a heavy burden and provided unequivocal support while I completed this book.

Finally, my first monograph is dedicated to the two most important people in my life, DLSS and KJSS.

PART I

BUTCHERY AS CRAFT AND SOCIAL PRAXIS

CHAPTER ONE

ANIMAL BODIES, HUMAN TECHNOLOGY

FROM FLESH TO MEAT

Some 2.5 million years ago, the evidence points to a radical departure in how our early ancestors acquired flesh from the carcasses of animals. Archaic hominids began to mediate their consumption of meat through technology. By using lithic implements to butcher, a small step for these proto-humans initiated a gradual cascade of new ideas and practices, becoming a crescendo that would drastically alter how humans interacted with each other, with other animals, and with their environment. In one way or another, many of the ways in which humans interacted with animals ended at the edge of a knife, a knife driven by *knowledge* of butchery.

Butchering is a uniquely human characteristic. Butchery is a concept that does not find expression in the natural world, despite the fact that other animals 'dismember carcasses' (Lyman 1994: 294). Initially, at an early stage in evolution, humans employed simple techniques and basic technology. For an immense time span, it seems, relatively little changed. Skipping many millennia, as humans move beyond the initial stages of domestication and husbandry intensifies, the implications of butchery for wider social practice become more evident. From the mid-Holocene, perhaps as a consequence of the influence of agriculture, we start to see steady, and at times explosive, modification in both the technologies of butchery and in the ways the animals are processed. One of these explosive moments coincides with the advent of metal: the techniques and tools that are developed make it easier to overcome the constraints of the animal's skeletal structure. More complex and varied knives are forged and utilised. New paraphernalia, like butcher's blocks and meat hooks, are created to facilitate processing. The scale of activity intensifies such that this additional equipment becomes a necessity. The practitioners themselves become specialised, branching out and diversifying into slaughterers, butchers, tanners, and horn workers. The networks around those who process animal bodies expand and include farmers, drovers, blacksmiths, and blade smiths. Roles within each craft profession are gendered, hierarchically structured, and diverse, with some working part time, others full time, and some seasonally.

A host of supplemental activity flows as a consequence of feedback mechanisms but also as a result of the diversification of the socioeconomic contexts of meat consumption. Butchery facilitates the scales of sharing and exchange, from interpersonal to long-distance transport of meat and by-products. Animal bodies are modified to accommodate the resources extracted from them. Initially, they are improved to increase their capacity for work, for traction in fields and for transport (Albarella et al. 2008). Later, changes in morphology become attuned to producing better-quality meat in greater quantity (MacGregor 2012: 426). With augmented production for flesh and amplified consumption, processing also intensifies, leading to special requirements to deal with waste, which has environmental ramifications. Originally, this is managed by simply siting carcass processors close to water (Goldberg 1992: 64–6; Yeomans 2007: 104–5). By the post-medieval period, in many large cities in Europe and America, processing is centralised and localised in the abattoir: geographies of slaughter become institutionalised (Lee 2008: 4).

The above describes the path of what is perhaps the earliest example of production and consumption, two perennial research topics in archaeology. Butchery complicates the connection between *making* and *consuming*. The butcher deconstructs a product of hunting or farming, the animal, and produces another, meat for cuisine. Thus, the activity of butchery occupies an interesting conceptual and intellectual space between production and consumption; it also mediates between animal body, food, and symbols.

A number of important questions – and concomitant hypotheses – emerge. For the prehistoric context, did changes in butchery also depend on a deepening understanding of animal morphology and ethology, at the point of transition from carcass disarticulation to true butchery? When did the division of flesh at a kill become sharing of meat in a settlement? For later periods, how do peoples' perceptions of animals change, for example, if we compare exploitation of wild versus domestic fauna, or with intensification in animal husbandry? Interpersonal relationships, within and between communities, affect how animals are perceived, hunted, and processed. What can butchery reveal about differences between categories of animals, systems of management, and the roles in society of those who slaughter, butcher, and process animal bodies?

The primary aim of this book is to examine butchery, butchering, and cut marks (an outcome of butchering) in layered, intertwined, and yet distinct relationships to one another. To do this, the book poses a key question: how do the operational sequences – the gestures, steps, and unfolding component parts – of a butchery event both reflect and shape wider economic and cultural drivers? What can the signs of butchery reveal about these broader conceptual and practical worlds in which the butchery took place?

The book picks up the story of butchery at the point when metal tools, specifically iron implements, become the mainstay of the craft, from late prehistory. Bone from this and later periods capture the dependence on agriculture and intensified animal husbandry, as well as increased centralisation of the population and burgeoning urbanisation. My emphasis on metal tools does not exclude those researchers working on lithic-tool butchery from this call for a far-reaching, root-and-branch revision of our collective protocols and methodology. The point is to observe and respond to the differences between assemblages created with stone versus metal tools in order to better understand how best to study each dataset. The approach and conceptualisation developed in the book applies to the spectrum of butchery studies. The introduction of metal is itself an important feature of changes happening at a societal level. The shift from lithics to metal knives for carcass processing represents the single most important development to have taken place in 'butchery' – as a concept and activity – until the advent of mechanisation.

The book also provides a methodological treatment of butchery. Studying archaeological butchery invariably involves the analysis of cut marks. These might be thought of as an indication of a specific activity, namely, the disarticulation of limbs and cutting of meat. But, as the arguments in this book will make clear, such a definition is too simplistic to describe *butchery*. Butchery involves cutting up animal bodies using tools according to a preconceived plan. Consider the physicality that exemplifies butchery (the activity), the intangibility of butchering (the cognition), and the progressive nature of the act of carcass processing. Butchery includes all of these components; as such, butchery data represent complex systems of interactions involving tool, practitioner, and carcass (Seetah 2008), and social and economic drivers (Seetah 2004, 2007). That such a complex constellation of interrelated factors would necessarily involve significant empirical variability should be clear, as should the utility of this variability for hypothesis building and interpretational breadth. However, in much of the literature so far, the tendency has been to focus on relatively narrow and specific aspects of the butchery record, which has hampered the scope of inference.

Gaps in our current approach exist for various reasons and stem from basic principles to do with how analysts approach the data. Butchery and butchering are situated at the intersection between the biology of the animal, the production and use of tools, and the cognitive expression of human intelligence and resource extraction. Zooarchaeologists, who are ideally situated to serve as the point of intersection, approach the topic from the perspective of one data source, bone. Though essential, by definition this is limiting. Cut marks are found on animal bone but are not part of the animal's biology. A conceptual incompatibility exists as the actual situation in life deals with flesh, with meat.

From a methodological perspective, analysts have yet to satisfactorily resolve a problem identified over three decades ago: 'an over emphasis on the minutiae of the cut mark' that derives from a focus dominated by bone (Dobney et al. 1996). To further complicate matters, methods to study cut marks have typically been developed from assemblages created with lithic tools, and then generalised for application across regions and time periods. Despite the fact that there are considerably larger, better-preserved faunal assemblages from historic periods, we still lack a dedicated recording system for metal-tool butchery. Where theory is concerned, we have not yet conceptualised, indeed intellectualised, what butchery represents beyond 'the removal of meat' (Russell 1987: 386), and the multifaceted role of the practice in society.

This book argues that it is now time to consider the limitations of these approaches and to begin to take up the task of improving on them. It does so by raising and examining a number of key points. First, there are fundamental differences between butchery using lithics versus metal implements (Maltby 1985a, 1989). Assemblages created with metal versus lithic tools are different in a number of significant ways. At the very least, we need to consider the utility of the more diverse cut marks that derive from historic periods, and how these might inform our methods. Invariably, distinctions also exist in how we interpret: meat is part of the process of 'calorification' or of 'commodification', depending on whether we are discussing prehistory or later periods, respectively. Finally, our recording systems do not easily accommodate the underlying fact that butchery, as activity, represents a trinity of evidence: the locational and typological characteristics of the mark, details of the tool used, and function. By recording only a portion of the data from the butchery record, it is difficult for analysts to infer on knowledge, intent, or cultural traditions.

However, perhaps the main barrier to overcome is that academic studies are absorbed with the products and outcomes of butchery. Anthropologists, historians, sociologists, and ethnographers discuss meat, the meat trade, and meat sharing and consumption. Zooarchaeologists study cut marks. Therefore, scholars interested in topics such as food assembly and consumption, and the place of animals in society, would benefit from a more nuanced assessment of the techniques and craft of butchery, the knowledge inherent in that practice, and the butcher.

PRACTICE

'Practice' and 'knowledge' serve as bridging agents, providing the impetus for employing analogy and ethnography, developing ethnoarchaeology, and undertaking actualistic studies – all approaches unified under middle-range theory. Indeed, through middle-range research, butchery has enriched archaeology (see Binford 1978, *Nunamiut Ethnoarchaeology*). Situating craftspeople has also been a concern to analysts using the *chaîne opératoire* approach, who endeavour to better understand activity from a range of perspectives (see Chapter 3). *Chaîne opératoire* helps to contextualise how knowledge and practice coalesce to reveal 'the whole person, indissolubly body and mind, in a richly structured environment' (Ingold 2000: xvii).

The constellations of activity involved in butchery, in combination with the ethnographic context, have provided the stimulus for my focus on practice. This, in turn, has been motivated by personal experience. My own background is in commercial butchery, trained for seven years in Brixton, South London. In addition to my time spent as a professional butcher, I have undertaken a range of other carcass-processing roles. These include traditional pig slaughtering and butchery, termed *koline*, in Slovenia, as well as knackering horses for the hounds of the Thurlow Hunt, Cambridge. I have prepared a wide range of animals, from llamas to wolves, for various zooarchaeological reference collections. As developed in Chapter 2, all of these experiences were episodes of 'butchery', the activity, but my actions and the drivers made each of these examples a different case of 'butchering': the behaviour, a cognitive process.

Thus, in this regard, *practice* refers to observable behaviours and sequences of operations undertaken in production. The activity that underpins practice relies on *knowledge*, which refers to a range of conceptualisation with which craftspeople engage (Keller & Keller 1996: 115–16). Craft is considered to be the actions of production driven by purpose and reason. It is utilitarian, holistic, and 'involves a rediscovery of subjugated knowledge, the recovery of practices made marginal in the rational organisation of productive routines. *The potter at the wheel must conceptualise the form desired even while pulling that form up from the lump of clay*' (Shanks & McGuire 1996: 78, emphasis added). Building on approaches to situate craft knowledge as an interpretative mechanism (Bleed 2008; Marchand 2010; Walls 2016), this book also integrates a range of perspectives garnered from the contemporary context. The practicing butcher does not see butchery in the same way as the faunal analyst and vice versa, and yet the differences between points of view need to be moderated if we are to conceptualise butchery. As a 'butcher', when I have recorded cut marks I have in my mind unspoken gestures of the body, inferring the movement of the hand and tool in relation to the carcass. I constantly shift between scales, going through a process of reflexive negotiation between archaeological bone and entire carcass. I effectively deconstruct the animal's body through a series of plausible possibilities to the point where I arrive at the cut mark I have in my hand; however, I am now armed with an explanation of the sequence of gestures, the steps, to produce the mark.

We cannot ignore economic and social drivers; indeed, situating these socioeconomic and environmental dimensions forms a major part of this book. However, viewed from the perspective I describe above, butchery is actually about knowledge driving the body and action. These actions define individuals and groups. In much the same way that learning the alphabet can lead to reading Shakespeare, so too can learning how to process a carcass lead to perceptual changes in the way humans view their environment, the animals within it, other members of their social group, and members of other groups (Yellen 1977; Binford 1978, 1981; Testart 1987; Kent 1993; Valerie 2000; Politis & Saunders 2002; Gravina et al. 2012).

Observing the performances of butchery from an archaeological context is not an easy task, even though the residue has been left on millions of artefacts. Studying the 'techniques of the body' (Mauss 1973) as they relate to butchery, how the activity has changed through time, and whether this can be assessed in a systematic manner is critical to the analysis of cut marks. It is this largely hidden aspect of our knowledge that I believe holds the key to more effective exploitation of the butchery record, and our ability as archaeologists to better understand ancient human–animal interactions.

THIS BOOK IN CONTEXT

This book ambitiously sets out to reconceptualise what butchery 'is' for an academic audience. To do this, it provides a more holistic approach to the theoretical framework from which we study the practice of butchery. The stress on practice, and particularly social aspects of practice, serves two functions. It aligns a growing trend in social zooarchaeology, recognising the richness and diversity of the human–animal relationship, with 'craft' – observed as a usable framework for assessing production, use, and discard of objects. More importantly, it confronts some of the consequences of economic determinism and methodological constraints that have hampered butchery studies. We understand meat because it is ubiquitous. We study meat today from a nutritional and cuisine perspective, and attempt to see similarities in the past, for example, through meat cuts (see Chapter 8). Zooarchaeologists conceptualise human–animal relationships through bone but do not situate the practice that transforms flesh into meat. Many of the social features attributed to meat consumption are dependent on butchery and driven by butchering. The nutritional context of meat is only part of the equation; indeed, *meat* is only part of the carcass!

Incorporating ethnographic research into this book provides an updated view of how different groups around the world engage with all parts of animal bodies, building on a strong foundation of this type of research in archaeology (Binford 1978; Brain 1981; Yellen 1977). The ethnographic context also provides a window on how Western views of meat have lost an essential connection to animal bodies, the skills associated with carcass processing, and, perhaps most obviously, the act of slaughter. Ultimately, by illustrating the contours of relationships between butchery, butchering, and practice, I aim to unite the subject matter with larger issues such as social organisation, cultural transitions, and routes to specialisation.

In writing this book, I have drawn heavily on my past experience to bring experiential know-how to an academic audience. In terms of constructing a new conceptual premise from which we can revitalise the study of butchery, the book incorporates a wide literature. Butchery and butchering are too complex, too deeply integral to culture, *too fundamental to people*, to be studied from one point of view or in isolation from one another. Alongside research on cut marks, the book borrows from studies of food as culture, food procurement as culture (e.g., hunting), the literature on meat, and the impressive body of work on technology and craft in archaeological contexts.

In this way, I aim to illustrate the nuanced, complex, and rich position that the topic holds in society, rooted in everyday activity. In order to better illustrate daily practice, I engage with a range of ethnographic studies, some of which are based on my own work. However, as important as it is to illustrate diversity, in the interests of thoroughness and to provide detail I have limited the text in specific ways. To achieve a balance between breadth and depth, the extended case study – used to showcase some of the ways we can better interpret the archaeological record – is focused on Roman and medieval Britain. This provides boundaries for the chronological and spatial contexts. In addition, cattle are the main domestic animal discussed throughout the text, and in the archaeological case studies a focus on cattle serves as another boundary for the book.

The book is split into two parts. Part I introduces the archaeological context, then deliberately steps away from archaeology and engages with modern case studies to situate the craft, practice, craftspeople, and technology within the book's conceptual framework. By tackling some of the gaps in our approach to metal-tool butchery, and adding richness through analogy, Part I provides a new grounding from which to renew appreciation for the subject. Chapter 2 begins by conceptualising the main topics under review, offering definitions for cut marks, butchery, and butchering; the chapter then describes limitations in the current analytical process, identifying how this has hampered our ability to describe the actions' of ancient butchers. Chapter 3 positions the book within the wider theoretical discourse on 'activity' in archaeology, serving to marshal the ideas and concepts that have influenced the development of this book. Chapters 4–6 then develop the wider social and techno-logical contexts. These chapters are based on modern industrial and non-industrial case studies, drawn from published ethnographic accounts, the ethnoarchaeological literature, and my own ethnographic research, as well as an autoethnography from the modern trade. These chapters examine the 'practice' of butchery, and I deliberately deviate from a focus on cut marks in order to better do this.

Part II recentres the objective on archaeological enquiry. Equipped now with a more representative and accurate view of the craft and people involved, the book draws on Part I to illustrate gaps in our methodological approach to cut mark recording. This section of the book shows how to mitigate some of the challenges faced by analysts recording this complex dataset and the ways in which we can enrich our interpretation.

Chapters 7–9 assess the state of the art in archaeology, and how we can enhance our current approach. Chapter 7 discusses how cut marks have been studied from archaeological bone; Chapter 8 describes some of the negative implications for interpretation that have arisen as a result of limitations in our methods. Methodological problems are based on a simple premise: we have been 'observing rather than understanding'; we record marks, less often do we deduce practice. Chapter 9 offers a synopsis of a new methodological approach that places stress on process – the steps and organisation of butchery – as a way to overcome an overemphasis on the mark. The amendments I advocate are based on the principle of assessing the *process* of butchery, recognised as a key element of the practice (Binford 1978: 63; Lyman 1987: 252) but not utilised as a means of situating the craft. In this way, the recording system effectively encourages the analyst to build interpretation during data collection. The application of the approach proposed in Chapter 9 is explored in an extended case study in Chapters 10 and 11, which summarise and discuss the results from six British sites.

As a case focused on butchery, the book highlights issues that are relevant to archaeology. From a methodological perspective, we need to consider how the recording of cut marks can be adapted to better assess butchery, but also to become systematic and standardised, in other words, to make better use of archaeological assemblages. Zooarchaeology is increasingly turning to molecu-lar methods, which have been a boon for the discipline (Guiry et al. 2015; Hagelberg et al. 2015). However, while offering many benefits, molecular techniques are usually possible on only a small subset of materials and provide a specific type of evidence. Butchery connects us back to the materials, is low cost and widely accessible (with training), and provides evidence of a range of activities that cover both social and economic factors. Studying butchered faunal assemblages does not mandate specialist equipment, nor does it need to incur additional analytical costs. As such, it is accessible to the wider archaeological community. Enhancing our studies of butchery to include new enthnoarchaeological approaches (Chapter 4) expands the types of studies that faunal analysts participate in and their research outputs.

In concluding this introductory chapter, I want to emphasise the uniqueness of the butchery record, which acts like amber, capturing a nuanced indication of human activity and behaviour at a given moment in time. As I explore in this book, butchery provides a way for us to view a long list of human thought processes, spanning the mechanisms of social stratification, to the commodification of animal bodies, and ultimately, the transformation of those same bodies into the metaphorically powerful domain of meat.

CHAPTER TWO

CONCEPTUALISING 'BUTCHERY'

INTRODUCTION

Butchery, conceptualised as craft and practice, has significant potential for revealing a more nuanced assessment of the human–animal bond our fascination with meat as food and symbol, and the wider ecological implications of carcass processing. Before we are able to revitalise the way we study butchery, it is useful to first identify ways to address long-standing gaps in the literature, including the lack of a clear definition of the subject and an over-emphasis on 'cut marks'. Thus, starting at the most fundamental level, this chapter offers a reassessment and recasting of terms, to better define distinct and foundational aspects of the subject matter. No clear and usable definitions exist that differentiate between cut marks, butchery, and butchering. It is therefore worth carving out the space to conceptualise the differences between an outcome of processing, the activity, and the cognition. Following this, as a way to move our methodological approach forward, I outline why the butchery process could be adopted as a usable thematic framework for the study of cut marks. These cues are then picked up in Part II, where I use the evidence of butchery to examine how inherited shortcomings in our analytical framework have hampered our ability to better understand the place of animals in the past.

The remainder of Part I therefore deliberately steps away from the archaeological context and engages with modern case studies to situate the craft, practice, and craftspeople within our conceptual framework. By tackling some of the gaps in our approach to metal-tool butchery, and adding richness through analogy, Part I endeavours to reconceptualise the subject from its very foundation.

'TERMS' OF ENGAGEMENT

Defining a subject can be constraining, and this is not my intention. Rather, the purpose here is to offer parameters within which future research can flourish. Defining these terms has a number of benefits. At present, inter-changeable use of language renders interpretation imprecise and complicates comparisons between sites. Thus, the following offers a way to make the use of descriptive terms systematic, clarifies the distinction between allied terms, and emphasises the complexity attached to each individual topic. Ultimately, these definitions should reinforce the imperative for standardised terminology (Seetah in prep a). As these terms are already widely used, I adopt the approach of constraining some of the breadth of use for each. In addition to promoting standardisation, the following brings the use of specific terms in alignment with a broader conceptualisation of the subject matter, as craft and practice. I have borrowed heavily from other analysts, in particular the influential work of Binford (1978: 50–60) and Lyman (1994: 294–315).

Let us begin with terms that have already been demarcated but not concret-ised: *cut marks* are the residue, or signature, of butchery; they are analytical units (Lyman 1994: 303–4). *Butchering* serves as the descriptive term for the specific rationalisations that underpin the activity, based on Binford's description of 'butchering' as expressing the 'cognitive ideology of the actors' (Binford 1984: 245). These two terminologies are at opposite ends of the butchery in terms of complexity: cut marks are an outcome of butchering behaviour. Casting the net further, one definition provided by the OED for the verb *to butcher* is 'to cut up or divide (an animal or flesh) after the manner of a butcher' (Pearsall 2016). This definition serves well the purposes of this book, as it describes the actions of an individual skilled in undertaking a specific kind of task. It also serves to differentiate between unskilled activities, as might be found for example in modern-day households, and the skilled practices of a *butcher*.

The activity involved is complicated to define, particularly as here I would like to refer to processes and actions, rather than to a catchall expression. For zooarchaeological purposes, the activity – 'butchery' (a logical descriptor) – has not been explicitly defined. This is not to imply that definitions do not exist in the literature; the issue is more that 'the term *butchering* tends to hold different connotations for different analysts' (Lyman 1994: 294). Definitions range from the very short and succinct, e.g. butchery as 'the removal of meat' (Russell 1987: 386), a mechanical utilitarian description, to more in-depth appraisals such as Lyman's 'the human reduction and modification of an animal carcass into consumable parts' (1987: 252). If viewed from the perspective of a wider conceptualisation of the topic, these definitions conflate 'butchery' and 'butchering'. In both cases, two key aspects are missing from these definitions. First, there is no reference to the cognitive component of the butchery process (although Lyman refers to this indirectly with 'human reduction and modification'): what butchery *is*, a meaningful practice for humans performing it, as well as its outcomes, the culturally encoded 'plan of action'. Second, and equally importantly, there is no reference to tools.

As a starting point for a more effective and useful definition, the practice and craft of butchery can be viewed as a concept, one that is distinct from the act of carcass dismemberment. Carcass dismemberment, itself a complex topic, is by analogy a by-product of consumption. It is an action, not specifically related to or performed solely by humans. We can see animal flesh as analogous to carcass dismemberment, while meat corresponds to butchering. The distinction separates the naturally occurring resource and its exploitation from that which is created by human cognition: only humans butcher (Lyman 1994: 295).[1] In a conceptual form, therefore, *butchery* includes more than the physical acts of dismemberment, but encompasses and depends on culturally specific attitudes that guide the timing and actions involved. Thus, a definition of butchery needs to reference the tiers of complexity that influence the decision-making process, not only the individual cut marks: the principles and embodied techniques denoting *why* the actions were performed.

Binford emphasised this aspect in his research and explained butchery in terms of stages of activity, rather than a single action: 'through it the anatomy of a large animal is partitioned into sets of bones that may be abandoned, transported, or allocated to different uses' (Binford 1978: 63; see also Lyman 1987: 252).

However, to return to an important point made above, these definitions overlook the role played by implements. To achieve a more thorough definition of butchery, the relationship between butchering (the behaviour) and cutting tools has to be included with the conceptual framework described above. Tools, in this regard, indicate any implement fashioned and created for the express purpose of butchery and can include the ad hoc selection of expedient objects, from general-purpose knives to task-specific cleavers. Butchery in its conceptual form only begins when we see evidence of tools created specifically for this activity. Prior to the appearance of hand-axes, flakes, and other archaic lithic tools, we are observing carcass dismemberment in the archaeological record, not butchery.

The presence of tools provides the hard evidence for the necessary cognitive developments that accompany and define our observations as carcass processing (butchery), as opposed to carcass dismemberment. Implements exemplify the full breadth of activity, cognitive and practical, involved in the butchery process (see Fig. 3.3, Chapter 3). Situating cutting tools in this way also draws attention to their influences on the practitioner, i.e., how they facilitate the activity. Tools provide evidence for the depth of complexity, the thought processes, and the duration of the individual tasks that guide the practice of butchery.

I thus propose a usable updated definition of butchery as follows: 'butchery is the range of processes, employing implements, by which humans are able to disarticulate a carcass into units depending on ultimate use'.

The term 'range of processes' resembles the 'series of acts' highlighted by Binford (1978: 63) and Lyman (1987: 252), with subtle but important differences. A series of acts implies individual actions; in contrast, a 'range of processes' incorporates not only the individual cuts but also the tiers of complexity inherent in the decision-making process of the butcher. We can therefore incorporate the cuts, the techniques, and the principles together. 'By which humans' indicates a planned and cogently thought-out sequence, and 'disarticulate a carcass into units to a preconceived plan' brings in functionality, as well as the socioeconomic drivers for carcass processing. Inclusion of the term 'implements' deals with the omission in previous descriptions of butchery to emphasize tool use.

In summary: *butchery* is a process, an activity, legible in the collection of cut marks; *cut marks* are the observed artefacts of human action and intentionality; and, finally, *butchering* refers to that intentionality, the underlying cognitive idea, the behaviour. If we accept that at the most basic level of investigation we are undertaking 'cut mark analysis', then the subject matter for zooarchaeological purposes should be 'butchering studies' as this is the level to which we aspire. This final point situates the subject within a cognitive framework and emphasises the breadth of acts involved, as well as the individual vignettes that we need to view in order to envisage the entire picture.

EMPHASIZING HUMAN COGNITION

Establishing parameters within which we can better define the subject matter allows us to pose questions from a more structured vantage point. What type of knowledge did the human actor operate with, and which we are attempting to study? Herein lies the most important feature of butchery: we have insight into the cognitive drivers involved. I argue that the collection of human practices, which we can consider to be butchery, involve particular kinds of cognition (butchering) that are distinct from other types of human–animal activity. To butcher a carcass requires forethought and intentionality, serving to separate it from other signatures of human actions that are left on animal bone. This is not to suggest that burning or fragmentation, for example, are not important markers of cognitive behaviour, but rather to acknowledge the breadth of activity involved in and around butchering, and to argue that cut marks may become a heuristically more important source of evidence of cognition than other taphonomies.

Efremov's (1940) original conceptualization of taphonomy (the study of how artefacts and ecofacts become part of the lithosphere) included cultural processes. Indeed, the principles of taphonomy (how bias is introduced or information is lost) are essential for the study of cut marks (Lyman 2010; see Chapter 7). Nevertheless, relying on taphonomy as a framing device is limiting because the taphonomic lens has tended to cause analysts to focus on distinguishing between mark typology, rather than emphasising the cognitive basis for butchery (see Chapter 8 for more on this topic). Cut mark data can provide a direct insight into the intentionality of individuals and societies. However, a problem arises in how taphonomic principles have been applied in zooarchaeology, as an approach for studying the processes of 'information loss or distortion'. Focusing on processes of loss or distortion has led to an emphasis on the geometry of marks, and how cut marks differ from other forms of abrasion (Pineda et al. 2014). This form of taphonomic enquiry is important but is less useful to those analysts who study metal-tool butchery marks. In the majority of these cases there is little, if any, doubt that the modification is in fact a cut mark.

The cognitive approach has been developed almost exclusively for application to studying early human evolution. Lithic implements, and the cognitive processes that underpin their production, have been a major focal point for these studies (see Chapter 7). The functional aspects of lithic tool use cover the gamut of non-utilitarian aspects, such as symmetry (Gamble 1998; Kohn & Mithen 1999) and butchery (Kleindienst & Keller 1976; Machin et al. 2005; Lapham et al. 2013: 141) and the complex cognitive behaviours that underpin biface production and use (Brenet et al. 2017).

'The human socio-cognitive niche' has been developed explicitly for understanding the evolution of human cognition (Whiten & Erdal 2012). However, despite such pertinent defining characteristics to butchering studies as *cooperation, egalitarianism*, and *mind-reading* (theory of mind), this approach does not engage with butchering as a potential mechanism to gain insight into the evolution of cognition. It seems extraordinary that something that is so obviously a complex expression of human behaviour has not been more effectively mined to investigate cognition. Butchering studies would no doubt gain considerably from being united to, and assessed from within, this theoretical vantage point. In describing what makes humans *human*, DeVore and Tooby emphasise the capacity for 'conceptually abstracting from a situation a model of what manipulations are necessary to achieve proximate goals that correlate with fitness' (DeVore & Tooby 1987: 209). This statement captures the essence of what butchering demands in practice: the ability to abstract, and from this, to construct a concrete plan of action to achieve desired goals.

The extensive literature on lithic tools illustrates the power of adopting a cognitive approach (see Lyman 1987). Unfortunately, butchery with metal implements has embraced little from this framework, despite potential benefit for a better understanding of reasoning, a topic I discuss in more detail in Chapters 3 and 4 (covering ideology in craft production). For metal-tool butchery, 'cognition' and behaviour have different meaning than they do for studies focused on early human evolution. Cognition, for zooarchaeologists working on faunal assemblages created using metal tools, relates to specific drivers. Some of these are behavioural at the societal level, not at the individual, for example, intensification in trade. Therefore, it is more appropriate to equate 'behaviour' with 'drivers' to avoid confusion. From this vantage point we can begin the process of rethinking metal-tool butchery by posing the following questions:

- What were the demands *on* butchery *by* society? In posing this question we can appreciate that butchery is more than an activity undertaken to exploit primary resources.
- What was the role of the butcher in planning and performing the task of butchery, and potentially acquiring the raw materials for and/or producing the necessary implements? Resolving this question helps to position the butcher as a skilled practitioner, a craftsperson.
- How can we place the cognition of the butcher into context? The practitioner's thought processes are essential to understanding what butchery actually entails. This starts at the point when the carcass is acquired (or even before, if the individual butchering the carcass is also responsible for hunting and/or killing the animal), up to the point that the desired resources have been obtained.
- What were the wider interactions, for example, between the butcher and the tanner, horn and bone worker, or blacksmith? Within this category are issues of demand, methods of transport (of live animals, meat, bone, and hides), society's view on the butcher and butchery, and an understanding of the overriding implications different traditions of butchery had on the society in question.

These questions reinforce the argument that neither butchery nor animal bone can be analysed solely in a quantitative manner; in both cases a quantitative focus can limit the broader implications of the data (Orton 2012). This does raise the issue of replicability of qualitative results. However, for the case under consideration, i.e., historic bone assemblages, there is a wealth of text and images to support qualitative interpretations. Furthermore, as will be outlined in the extended case study, utilizing a methodological approach that integrates both quantitative and qualitative data provides a strong basis for wider interpretation.

The taphonomic perspective has been particularly beneficial in driving a specific branch of analysis within butchering studies (discussed in more detail in Chapter 7). However, it may not be the best approach from which to study the behaviours for which the cut marks are evidence (see Speth 1991: 37, who notes this as a more general concern in studies of taphonomy). Combining taphonomic and cognitive methodological frameworks would be mutually beneficial, as we pinpoint the details of the cut mark, while maximising our likelihood of understanding aspects of behaviour. Furthermore, to identify and bolster our most heuristically important dataset, the methodological approach should be driven by cognitive rather than taphonomic concerns; we are after all more interested in human agency than depositional processes and outcomes (unless these latter aspects also describe agency).

But how do we move from the cut mark, as taphonomic signature, to cognitive inference? Methodological tools such as ethnoarchaeology and experimentation have considerable utility in bridging these two ends of the butchery record (Chapter 3). However, there is a need to exercise caution when employing ethnographies. Using examples of early behavioural studies by Gould (1967) and Binford (1978, 1984), a synopsis of the debate can be summarised in a question: could the butchery pattern of the Ngatatjara from Australia be described as rigid, compared with that of the Nunamiut Eskimo from Alaska? Distilling the main features of this debate, complex arguments were presented for why the individual patterns of butchery were observed, with the conclusion reached that the Ngatatjara showed a relatively fixed pattern of butchery, compared with the dynamic one of the Nunamiut (depending on the viewpoint of the analyst). However, in these studies, much of the debate was presented without recourse to the technological, cultural, or functional place of *butchering* within the wider community being analysed. To the extent that cultural attitudes featured in the discussion, they focused on sharing versus storage of meat, not the processes of butchery itself.

Butchery is conservative in terms of variation, compared with stylistic and artistic attributes that can be incorporated into the production of ceramics, for example. The skeletal morphology of the animal serves both as an inherent mechanism of constraint and as a 'blueprint of expectations'. Patterning in butchery is dependent on tools, cuisine, and socioeconomic drivers. One needs to understand the details of the implements, how they were manufactured, and their maintenance, in order to appreciate any significance the tool may or may not have for the group in question. Thus, ethnographies have an important contribution to make in terms of understanding how we can move from patterning of marks to cognition. However, if we are to truly take advantage of the potential of ethnographies, the scope of research must emphasise the place of butchery – and indeed of meat more generally – within society. For example, are feasting practices guided by patterns of meat sharing? Do certain individuals receive specific portions? Do these in turn guide the process of butchery?

CONCEPTUALIZING THE 'WHEN' AND THE 'WHAT'

An established concept in *chaîne opératoire*, an approach developed to assess the sequential nature of production and use of objects (but here is applied to food, see Miracle 2002: 67) has been the idea that isolating evidence of technology from the material record depends on an 'inferential procedure' (Sellet 1993; see Chapter 3 for detailed discussion of *chaîne opératoire*). The notion of building inference during the recording phase is structural to the approach to cut mark analysis advocated in this book and will be developed in Chapter 9. However, here I would like to introduce the reader to the principle that underpins the development of this specific methodological approach. The following also situates how the knowledge of butchery might be conceived of from the faunal record, providing insight into the data the analyst is interested in capturing.

'Amber Moments'

Cut marks are the preserved signature of an action, manifested in a form that retains specific details of a unique snapshot in time: metaphorical amber or, more precisely, an *amber moment*. This idea emphasises the ellipsis that exists between the study of cut marks and the interpretation of butchering behaviours. What are we actually analysing, and how does this relate to the way we infer on the utilisation of animal parts in the past? By creating a gap in terms of when analysts record marks, and *then* start interpretation, we miss the essential *moment* of capturing the true essence of the butchery record. The significance of these two moments, when we record, and when we interpret, in this case, is as important as what we are recording. This is because, once we remove the mark from our sight, we are attempting to interpret from an imagined space. Naturally, we can come back to the *moment* if we have access to the assemblage; however, in practice, we almost universally interpret from a distance. Such moments are legion in archaeology, for example, when observing a fingerprint in a clay object or the striking platform of a flint tool, or recording the rising of the sun in order to position an ecclesiastic structure. In each of these cases the concept has different implications.

In the contemporary situation, cut marks are often recorded as an alphanumeric string code, describing location, type, and number of marks (Fig. 2.1). These tend to be relatively conservative in the diversity of marks they can capture (without significant annotation) and are often unique to an analyst. In response, more recent attempts have been made to record marks topographically, using high-resolution templates of skeletal drawings (Fig. 2.1). In both recording

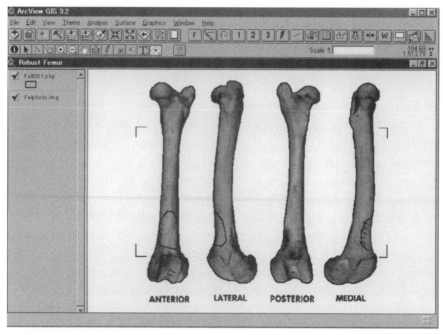

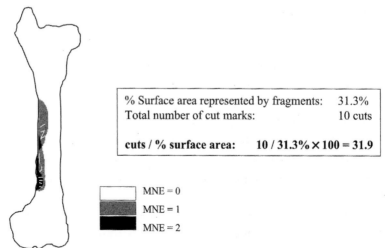

FIG. 2.1 Template recording system, using topographical markers to note type and location of marks (input screen above, cut marked record below). (From Abe et al. 2002. The analysis of cutmarks on archaeofauna. *American Antiquity* 67: 6443–63. Reproduced with permission.)

scenarios, a lag exists between recording and inference building that potentially hampers deeper interpretation (discussed in detail in Chapters 7 and 8).

Thus, in practical terms, applying the concept of 'amber moments' for recording butchery may help close the gap between recording the specifics of location and type and interpreting the cognition. From a theoretical perspective, the focus shifts from the cut mark to the behaviour, at the point of analysis. It reveals the knowledge that is driving the butcher and moves emphasis away from noting the geometry of the mark or the number of marks present. In this way, we make the process of interpretation more precise and streamlined, which could potentially remove a number of filters, for example:

- Time: as it relates to the period between recording the mark and interpreting the behaviour.
- Coding: which bears no resemblance to either the mark or the behaviour and serves primarily as an academic response to chronicling practical knowledge. Reinterpreting a string code after the fact inserts yet another filter, further increasing the gap between cut mark and behaviour.
- Template/anatomical recording: undertaken to assist with and collate the position of the marks to allow for later interpretation. However, representing *all* marks from an assemblage on a diagram of one bone/animal is a fabrication of a butchery 'pattern' that no single animal, or body part, underwent.

Removing filters does not imply rejecting these approaches; coding and templates are structural elements of our recording and offer many benefits. The point is that each introduces a specific bias that can be mitigated. By modifying how we use these recording principles and introducing 'process' as the motivation for what we record, we are able to capture a more complete dataset. From this conceptual basis, we are in a more effective position to move toward standardisation in recording the marks and improving replicability. Standardisation in this case does not refer to using the same string code as other analysts, which serves only to compound and repeat an already problematic situation. Standardisation here implies a methodological basis built on a set of parameters that are intrinsically and irrevocably tied to the specific subject matter, in this case, the morphology of the animal, the characteristics of the tool, and the relative positions of the practitioner and carcass during butchery.

It is impossible to interpret the totality of these factors from the cut marks, and that is neither my implication nor aim. However, the collective influence of these factors is important. Consider the impact of these multifaceted areas of knowledge and the implications they have on our methodological protocols and inferential framework. Cut marks capture these varied influences in a single snap-shot of time, an amber moment. It is for this reason that timing of interpretation is critical: detaching ourselves from the moment serves to reduce the accuracy and depth of interpretation.

Butchery as Process

We must develop not only better theory for conceptualizing processes but also more adequate methods for studying them.

(Vogt 1960: 400)

The processes Vogt referred to more than half a century ago were cultural, broadly defined. Though deliberately taken out of context, in many ways the above quote sums up precisely what is required for better use of cut mark data. As Lyman emphatically states, 'Butchering is process' (Lyman 1987: 252), an idea long since echoed by other analysts (Guilday et al. 1962: 59; Binford 1978: 63; Greenfield 2016: 89). However, to my knowledge, the analytical framework advocated in this book (Part II) is the first instance where the inherent process of butchery – the aspect that describes the transformation from flesh to the production of meat itself – has been proposed as a usable framework for cut mark recording. This stance derives from a desire to situate the craft and craftsperson. The following discussion focuses on how we can actually assess the 'process of butchery' and leverage evidence of behaviour. It raises an important question: can we meet the challenge of interpreting *process* at the same time as *recording* cut marks? Not only would this potentially be a more efficient approach to cut mark analysis, it also directly influences our ability to decipher and interpret cognitive drivers. The types of information butchery data could potentially reveal have to be properly situated, and one way to do this is to consider the diversity of knowledge bases and socioeconomic drivers that influence this craft (Chapters 2–6). This standpoint provides the grounding from which to develop and utilise a recording method that guides us to capture process from cut marks (elaborated in Chapter 9).

What is meant by process in this instance? If the reader is willing to accept that the mark is the outcome, then in this context, *process* refers to the route through which a carcass was butchered (Gifford-Gonzalez 1991; Fisher 1995). In the immediate sense, it means the trajectory of the cutting implement and the force used to create the cut. Indirectly, this also includes evidence of the relative position of the carcass, practitioner, and tool, and a network between people and things (Latour 2005; see Chapter 3). To move forward a system is needed that facilitates interpretation, one that is created not from the viewpoint of what we can see but from a new perspective emphasising sequences and procedures. Of course, we start from the tangible: the details of the marks are recorded, an essential step that cannot be circumvented. However, by conceiving of butchery as process it is possible to move 'beyond the minutiae', guided to assess and observe different features of the mark itself during data capture. The framework of the system utilises those features that constrain butchery: the animal's morphology, the suite of tools available for use, and the functional nature of the task. This last point includes the fact that there is a finite number of ways to disarticulate a carcass, despite the opportunity for cultural expression. In addition, the scheme is enhanced by having built-in ways to record those knowledge systems that drive the butchering: the technology (tools) and the behaviour (function).

The methodological component needs to start from a new precept: rather than catalogue, we need to *read* butchery. Like all literature, we gain under-standing not from an emphasis on individual letters, the cut marks, but by viewing the collective of sentences and paragraphs: the butchery process. But this is only the starting point. The idea of reading an artefact has a well-established precedent in archaeology (Hodder 1982). However, this notion has been critiqued in that it represents only a component of how we should consider and think about people in the past. As important as 'reading' the past is our ability to recognize the whole *schema* of bodily movements and gestures, the nuances of practice, what actually takes place during physical activity, particularly when objects are being created. Such ideas borrow heavily from phenomenology. The essential development I am proposing is that we embed this framework into our recording of cut marks. By situating activity, the knowledge inherent in production is brought to a 'reading' of the archaeo-logical record. As will be outlined in Chapter 3, gestures, movement, and embodied actions, as well as the relationship between operational sequences of production and bodily engagement, are perfectly suited to deepen our study of butchering. The principles of experimental archaeology, phenomenology, and the archaeology of the body are all theoretical positions that have much to contribute to studies of butchery.

Accurately assessing process relies on more than simply distinguishing between taphonomic signatures. The analyst must ask: can we deduce a rationale for why the marks were produced? Why are they in that specific location? What has caused them? Why and how were they made? It is also important to consider a number of underlying principles, for example: the marks were not made deliberately; the practitioner was concerned with maintaining a keen edge on the implement; the practitioner's physical effort was moderated, exploiting natural weaknesses of the animal's skeletal morph-ology to facilitate butchery. Identification also needs to be reflexive (Hodder 1997). If marks noted during data collection reveal the intended use of the meat, i.e., an indication of cuisine or the other commodities sought from the carcass, these factors should be taken into account during subsequent recording.

Chapter 9 deals specifically with the practical application of these principles. However, it is worth stating here that process is not an ephemeral criterion. Recording butchery as process is a relatively straightforward task that can be easily integrated into the analysis of faunal remains. Indeed, logging the details of butchery in this manner is intuitive. Butchery involves cutting up animals using tools to a preconceived plan. Thus, analysts are interested in recording the tools, the cuts, and their relative position on the carcass (the current level of detail); deducing and describing the functional aspects of the butchery then enhances the capacity for interpretation. In addition, for logistical reasons but also to facilitate notation, a simple way to describe the idiosyncrasies and nuances of individual marks has to be embedded into the recording system.

These parameters serve to guide and strengthen how we study cut marks. By placing butchering within technological and social frameworks, and recording process rather than outcome, we can make a novel contribution to zooarchaeological analysis of bone. Equally relevant, these principles may find resonance for the study of pots, glass, hand-axes, or figurines (Nakamura & Meskell 2009).

CONCLUSIONS

Butchery data offer a powerful tool for inferring cultural or economic characteristics of the human–animal relationship. In examining butchery data, the analyst confronts a host of important factors that complicate the interpretation of the data, e.g., the fact that the majority of cut marks are incidental; that the cut mark retains details of activity that is difficult to understand without specialist craft knowledge; and that any single mark could be part of a complex sequence of events that may well be staggered and involve the agency of numerous individuals. Viewed from this perspective, it becomes apparent that for the analyst to capture a more complete dataset, both taphonomic and cognitive approaches are necessary.

The proposition argued for in this chapter offers more than a review of previous descriptive paradigms. It provides an innovative analytical approach, which could be easily applied to a range of settings. Animal bones are often the most numerous ecofacts recovered from a site, and a significant proportion of these remains will have cut marks on them. However, we must look beyond this corpus of material, immense as it is. The next four chapters deliberately take the reader well beyond the 'cut mark' to position the importance of theory in situating the social context, the role of ideology as a driver of production and craft, the relationship between craftspeople and society, and the significance of the technological features of butchery. These chapters implicitly question the way ethnography has been utilised in the pursuit of knowledge of the ancient human–animal relationship, comparing this with cases taken from studies of other materials. In this way, the book makes the case for why we need to enhance and diversify our approach to faunal remains.

CHAPTER THREE

THINKING PRACTICALLY

INTRODUCTION

Traditional forms of pig slaughter, as opposed to commercial production and processing, have a long-established precedent in Europe and are for example depicted in Queen Mary's Psalter, dated to between 1310 and 1320. *Koline* describes the seasonal, traditional form of pig slaughter as practiced in Slovenia (Smerdel 2002: 35). In 2013, I participated in this festive occasion in Birčna Vas, a village 3 kilometres from Novo Mesto, a city in south-eastern Slovenia close to the Croatian border. During the festival, I undertook the role of butcher. The experience left a strong impression on me about the power of analogy, its uses in understanding ancient human–animal relationships, and in particular the utility of analogy in situating how human–animal relationships unfold in a given social-historical setting.

Koline takes place during cooler autumn weather, usually during November, and is localized around the home environment. The occasion adheres to traditional norms, as understood by participants, and generally eschewing the use of special equipment or large-scale refrigerated storage. The process starts early in the morning: water is first boiled, then used to scald and de-hair the pig carcass. The animal is slaughtered by an incision into its throat and the blood is collected for sausages. The remaining butchery involves cutting pork steaks and making sausages and *špek* (bacon). Part of the reason for undertaking the tasks outside is that the fatty tissue of the dead animal becomes somewhat solidified due to the cooler outdoor temperatures, which facilitates the process of making sausages.

Koline has been a feature of life in Slovenia since at least the sixteenth century (Novak 1970: 380). It articulates a division between rural and urban society, one focused on participation and networks of meat sharing. As an event understood to incorporate both the festive and the everyday, *koline* is a uniquely revealing case study for assessing the social and economic role of meat and processing. The term itself incorporates both the act of butchery and the collective products. 'The gift of *koline* [the products] is confined to a set of material and social relations in Slovene society which find their locus in those households which domesticate and slaughter pigs' (Minnich 1990: 152). Commercial pork can never be transformed into *koline*. If these points situate the social setting, the economic context can be understood by the fact that at least until the 1990s *koline* provided the majority of pork for the general populace; indeed, the circulation of pork from *koline* and commercial production remain inseparable from each other (Minnich 1990).

As I butchered the carcass that November morning in 2013, it was evident that the knowledge of butchery was indispensible for the task at hand and for the collective enterprise more broadly. The reason the event could be held 'at home' was because someone with the correct expertise was present. The alternative would have required the processing to be undertaken at a butcher's shop, which effectively breaks one of the ties with tradition, namely, that this event takes place around the homestead. In this instance, to satisfy veterinary legislation governing proper, humane, and hygienic practices, the pig was slaughtered on the farm where it had been raised. A licensed local butcher had travelled to the farm, and after slaughter, the carcass was transported to Birčna Vas for the remainder of the butchery.

As I had been forewarned about my involvement in this *koline*, I had taken steps to perform an assessment of the animal's usability based on body parts, i.e., to produce a 'pig utility index', following the well-established protocol laid down by Binford (1978: 12). The utility index provides a means of appraising the relationship between carcass uses, constitution, and transportation. It offered a means to assess how human behaviour acted as a driver for selection of animals and their parts. Whether employing Binford's model (the modified general utility index, MGUI), or derivatives, the method depends on scaling the percentages of meat, bone, and grease (Metcalfe & Jones 1988).

However, as I weighed the skin, meat, and bone, and then spent a day rendering the grease from the bone to weigh its fat content, I was struck by the incongruence between the episode of *koline* and my subsequent actions during the utility index study. In other words, the type of approach I was using was not particularly appropriate for investigating the relationship between the social features of *koline*, the practice of butchery, and this form of provisioning.

In fact, I observed closer correspondences between the utility index and commercial butchery, rather than *koline*.

Binford's work was never intended for application to a historical context such as *koline*, whether methodologically or in spirit. However, I use this episode to demonstrate the difference between the types of enquiry into human–animal interactions that captivate zooarchaeologists focused on different periods. This is particularly relevant when thinking about the scales of consumption examined by archaeologists working on later periods.

This chapter takes the view that new approaches, sensitive to the types of large, complex assemblages recovered from historic sites, would benefit research on human–animal relationships, food and cuisine, craft development, and trade. In particular, for studies of butchery, teasing out the nuances of activity through modern correlates and analogy could help illuminate broader social aspects of production and consumption. The following reviews a number of theoretical approaches that archaeologists have employed to analyse/understand/interpret the actions of past craftspeople. Using this broad overview as a point of departure, I then explore how these approaches have informed the development of a new conceptualisation of butchery advocated in this book, one grounded in a view of butchery as craft and practice. The chapter also emphasises the development of specialisations within butchery, while noting that this specialisation did not preclude interdependence with other allied crafts and professions. Thus, the picture that emerges is of a 'butchery ecosystem' that includes deep connections among humans, animals, objects, and environments. Ultimately, the emphasis on practice draws attention to the sensory facet of the craft: the application of force to flesh – the inherent response to the material – as the butcher turns that flesh to meat.

ARCHAEOLOGICAL THINKING ON ACTIVITY

Analysing 'butchery' as craft and *praxis* has mandated a complex negotiation between different theoretical positions. On the one hand, there are approaches developed specifically for the study of faunal remains, while on the other, there are a range of models that tackle 'activity' as it relates to physiology, psychology, society, and landscape.

In assessing the significant theoretical developments as they relate to cut mark analysis, Binford's work represents a significant milestone. While White is rightly credited with drawing a wide audience to appreciate the empirical potential of cut marks (White 1952), Binford's research was crucial in advancing a methodological position for *how* and *why* we should study this dataset.

Binford introduced the concept of middle-range theory to archaeology, adopting the framework from social studies (Binford 1977, 1983: 10). Middle-range theory provided analysts with the ability to build varied modern reference collections that could be used comparatively for assessing the past (Atici 2006). It stipulated that methods of inference, but not the outcomes, should be tested through analogy (Binford 1983: 12–14). As a model, middle-range theory was positioned between the general theories that drove the discipline and empirical studies: hence, 'middle-range'. Binford's approach aimed at bridging the gap between the perceived static archaeological context and the dynamic one of the modern day. To achieve this, middle-range theory depended heavily on experimentation and replication, analogy, and ethno-archaeology (see Chapters 3 and 4). I discuss replication and butchery more specifically in Chapter 8, but it is worth noting that studies range from the contemporary reproduction of artefacts (LeMoine 2002) to establishing the production sequence and apprenticeship strategies used to create commodities (Wallaert-Pêtre 2001). Experimental archaeology (reliant on hypotheses, usu-ally tested using replication to approximate an understanding of ancient technology or practice) has been important for understanding innovation and perpetuation of material technologies, the cultural and pedagogic influ-ences on the production of objects (Coles 1979), performed tasks such as knapping (Frison 1989), and aspects of human cognition (Isaac 1981; Bell 1994: 312–25). Middle-range theory also relied on the principles of taph-onomy, as method and theoretical framework, and uniformitarianism, e.g., the notion that natural forces will act in a similar fashion throughout time (Atici 2006). The utility index exemplified the principles of this theoretical model and has been adopted to develop indices for the horse (Outram & Rowley-Conwy 1998), porpoise (Savelle & Friesen 1996), seal (Savelle et al. 1996), bison (Brink 1997), kangaroo (O'Connell & Marshall 1989), and gazelle (Bar-Oz & Munro 2007), to name but a few examples.

Middle-range theory has not been without its critics (Tringham 1978; Raab & Goodyear 1984; Trigger 1995). One of the more fundamental concerns centres on the assumption of uniformitarianism (Gifford-Gonzalez 1991). Binford himself recognized that this principle presented a limiting factor when one considers the time depths with which archaeologists habitually work. He suggested that uniformitarianism was applicable to three areas: the use of space by people, the resulting spatial structure of discarded objects, and the ecology and anatomy of extant fauna (Binford 1977, 1981: 44). Approached from an historic context, however, none of these holds true. With centralised popula-tions and subsequent urbanisation, the use of space became highly nuanced, as did practices to deal with waste (O'Connor 2008: 160–1; Bartosiewicz 2009). Through selective breeding, we have modified the constitution of domestic animals to an extent that renders them distinct from their wild counterparts and unable to thrive without human intervention (Diamond 2002). For those studying historic assemblages, this type of approach has limited applicability considering the numbers of animals being transported, often on the hoof, and over long distances within a well-organised system. Furthermore, to return to the case of *koline*, this perspective fails to take sufficiently into account the complex web of social activity around processing and consumption.

Despite these caveats, Binford's work with the Nunamiut illustrated how replication could be used to test hypothesis and to generate analogy that could then be applied to archaeological questions (Mathieu 2002: 2). Based on this work, Binford observed that the pattern of discard following butchery was as a consequence of distal limb bones, with low meat yield, being left at the kill site rather than being transported back to the camp (Binford 1978). Similarly, Lyman (1994: 299) presents a conceptual framework for analysing butchery based on transport. Binford's and Lyman's work provide important examples of the conceptual framework that governs activity, whether viewed as 'economy' or 'transport'. Building on this, it is possible to use 'craft' or 'practice' in a similar fashion, as framing mechanisms. To do this, one must draw from a wider research base that is more specifically aimed at understanding the relationship between human actions and the production and use of objects.

Schiffer was fascinated with the centrality of human behaviour and its relationship with technology (Schiffer 1992). While Binford's approach was primarily empirically driven, Schiffer was concerned with the stages and steps involved in activity patterns, the residues these left, and how these might be connected to transitions in technology. Schiffer used a variety of analogies to illustrate his principle of 'behavioural archaeology'. Similarly, his work and that of his colleagues on the 'behavioural chain' complements and enhances *chaîne opératoire*, discussed below, through an emphasis on historical contextualization (Schiffer 1975, 2004; Skibo and Schiffer 2008: 21).

Where Schiffer's approach brings the sequence of production to the fore within a larger framework describing changes in technology, other theoretical models have sought to interrogate the social relationships between object and agent. Actor–network theory (ANT), for example, developed within science, technology, and society studies, refuses to prioritise either human or non-human agents, placing all aspects of a social situation on parity. Humans, non-humans (objects), and ideas are afforded the same value in terms of describing (not explaining) social interventions. Callon (1984), Law (1984), and Latour (1996, 1999, 2005) championed ANT, deemphasizing human subjects and adopting a symmetrical approach to the study of social phenomenon (Latour 1996; Knappett 2008: 140). This stance, focused on equalizing the various components of social interactions, has appealed to archaeologists keen to overcome dualisms: people versus artefacts or biology versus culture. In turn, ANT has given rise to 'symmetrical archaeology', characterized by the work of Olsen (2003), Shanks (2007), and Witmore (2007), for example. As a field of analysis, ANT is deeply connected to production and to materials, posing critical questions about how 'things and materiality at large relate to human beings and "social life'" (Olsen 2003: 87) and offering the suggestion that 'making things makes people' (Shanks 2007: 591).

In approaching butchery as practice and craft, middle-range theory has much to offer by way of conceptual frameworks, such as transport: how did humans conceive of moving parts of animals they butchered around the landscape? Similarly, Schiffer's ideas of sequence centralise the importance of 'stages' during the creation of products. ANT in turn places the social context (but not social power – see Preucel 2012) at the heart of production and suggests that interrelationships between components of any activity-mediated production process are critical: 'there is a clearly definable task at hand, which is to be achieved by one human in conjunction with one tool' (Knappett 2008: 140). ANT draws on a defined sense of materiality. This can take the form of selecting the correct test tube to conduct a successful scientific experiment (ANT was developed to study laboratory-based research, see Latour 1987; Van Oyen 2015) or, as might be applicable in the present case, the correct knife for a specific butchery task: a fileting knife for defleshing. However, where ANT has a concrete contribution to make for butchering studies is in drawing attention to the complexity of interaction, the networks that develop along and between the chains of production, use, assembly, and discard. Consider a butcher, commissioning a knife from a blacksmith, who in turn acquires horn from the horn-worker to produce the handle for the knife, which was sourced by the horn-worker as raw material from the butcher. As a consequence of specialisation, the objects and *actants* (as per Latour 2005: 54–5) remain intimately interdependent. This example illustrates a benefit of engaging with ANT, but also reveals a gap: the inherent egalitarianism of ANT leaves relations of social power underspecified (Preucel 2012). As discussed in Chapter 4, status and social stratification are often, if not always, critical to assessing the wider context of craft production and craftspeople (see also examples in Sterner & David 1991).

These models situate activity within technological and social settings. Ingold brings both the networks and the tasks together, under the umbrella of the *taskscape*: 'Every task takes its meaning from the position within an ensemble of tasks, performed in series or in parallel, and usually with many people working together' (Ingold 2000: 195).

The relevance of this notion of taskscapes for a butchery context became acutely clear to me during fieldwork in Kimana, Kenya, in March 2017. The purpose of the trip was two-fold. First, I went to Kimana to study the different tiers of slaughtering practices from homestead to abattoir. Second, I wanted to see if a relationship existed between slaughtering practices in different locations and the potential risk of contracting Rift Valley fever virus. The first and most basic tier of commercial slaughter, colloquially termed a 'slab', provides meat either for the local population or for immediate transport by road to the wider region, up to Nairobi, some 4–6 hours away.

The slab I visited had hanging paraphernalia to suspend carcasses, butchers' blocks, cleaning stations, and waste drainage systems in place. Groups of individuals worked at different tasks: corralling animals into the slaughtering block, slaughtering and butchery, cleaning intestines and other offal, or stripping flesh remnants from the recently flayed hides. There was little mixture between groups, but within 'task groups', individuals would switch between particular tasks, e.g., slaughtering an animal or hoisting it and starting the skinning process.

Timing was an important part of the decision-making process. For example, after being slaughtered, the animal was doused with cold water to encourage cooling of the body. It was then left for approximately 15–20 minutes to reach ambient temperature. During this period, the butcher would shift to hoisting another animal, and begin skinning. The tasks were not as defined as in the modern commercial context in Europe and the United States, or in the larger slaughterhouses and abattoirs elsewhere in Kenya. However, the relationship between groups and individuals, each interdependent on each other, not only presented a useful example of a defined taskscape but also highlighted the way tasks were mediated by factors such as time and space. Indeed, the individual and collective tasks orchestrated the 'system'. The person doing the slaughtering could not proceed until the previous carcass has been hoisted; the skinning and evisceration of that animal could not take place until the skin and intestines of the previous carcass had been removed to the next station; and so on (see, Fig. 3.1a–c).

As the above discussion has developed, the reader can appreciate the growing complexity of how archaeology has assessed 'activity', in broad terms. These models have provided the broader framework for moving butchery away from the immediate connection with cut marks per se, and instead to expand the focus to include the craft, social, and cognitive features of the activity. In this regard, and in addition to the models discussed so far, concepts derived from cognitive archaeology (Renfrew 1994: 3; Schlanger 1994: 143–5; Flannery & Marcus 1998: 35–49; Coolidge et al. 2015: 177–8) can usefully connect the behavioural features of butchering to an 'archaeology of the mind'.

Recently, building on the ANT approach (see in particular Callon 1984), the complexity that exemplifies the interconnections between people and things has led to new thinking on 'entanglement' (Hodder 2012). In many ways, the idea of entanglement describes the 'ecosystem model' for representing butchery as part of a series of coupled systems (see Fig. 3.3). There is also a growing interest in practice within social studies, which provides usable cases and perspectives built from contemporary enquiry into activity (Chaiklin & Lave 1993; Tobach 1995: 43; Keller & Keller 1996). However, while these latter examples represent concepts with considerable potential for future research on activity, they have had little influence on the study of crafts related to carcass processing. In contrast, one theoretical model could provide a clearer and more immediate contribution to the study of butchery itself: *chaîne opératoire*.

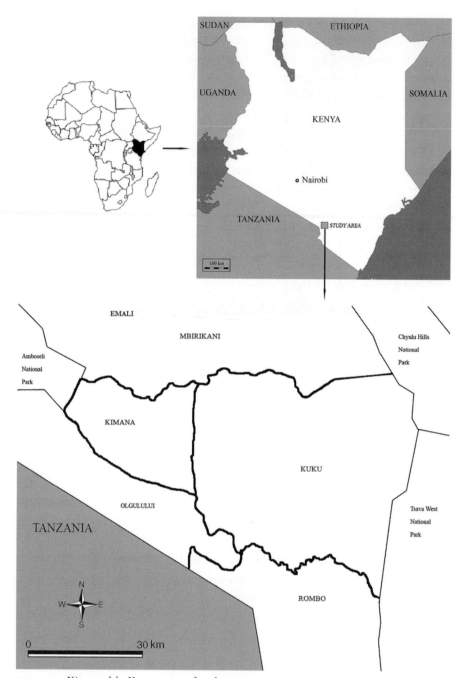

FIG. 3.1a Kimana slab, Kenya: map of study area.

FIG. 3.1b Kimana slab, Kenya: separation of taskscapes into skinning and butchery stations. (March 2017, by author.)

Chaîne Opératoire

The approach known as *chaîne opératoire*, which offers a powerful set of tools for looking at sequence of production, has traditionally been applied to research on European lithic technology. *Chaîne opératoire* refers to the 'incorporation of the process of production and use into classification and interpretation' (Andrefsky 2005: 38). It has been employed for other material cultures where processes of production and use have bearing on wider interpretation, such as studies of ceramics (Wallaert-Pêtre 2001). More recently, as *chaîne opératoire* has developed into a tool for dismantling typologies (Walls 2016), it has found application in a broader range of topics, from cognition to describing agency (Dobres 2000; Stout & Chaminade 2007).

Chaîne opératoire derives from the work of A. Leroi-Gourhan (Audouze 2002) and was adopted by French prehistorians (e.g., Geneste 1990). Similar schemes developed in other parts of Europe, the Near East, the United States (Bar-Yosef & Van Peer 2009), and Japan (Bleed 2008). *Chaîne opératoire* posits that a given instance of production is driven by a concept in the mind of the practitioner, as opposed to merely reflecting a mechanical reduction sequence (which for example exemplifies the American iteration, see Chazan 2009: 467).

At the core of the concept is the operationalised sequence. More recently, the emphasis on sequence (Sellet 1993) has given way to 'pattern' in order to accommodate for the fact that activity is predicated on both

FIG. 3.1C Kimana slab, Kenya: cleaning station. Though distinct, each grouping is interdependent due to finite space and time constraints. (March 2017, by author.)

sequence and structure: 'the acting out in time of knowledge and skill' (Chazan 2009: 468). From its naissance, *chaîne opératoire* has been concerned with understanding technology. However, in much the same way that 'sequence' has been enhanced to 'pattern', the idea that the technological chain ends at production has been superseded by the notion of an object's life history (Pelegrin 1990; Sellet 1993; Siller & Tite 2000). In this way, *chaîne opératoire* captures the stages from procurement of raw materials to production, use, and final discard: 'We need to study not only the origins of an object, but its entire life history' (Martinón-Torres 2002: 33). Indeed, when one considers the complex ways in which production processes and the object's biography are connected in order to better interpret the past, we have to acknowledge that *chaîne opératoire* also encompasses the work of the archaeologists themselves (see Meskell 2017 for a case study based on figurines and the body).

Chaîne opératoire has led to a range of methods that are instrumental in producing the data needed to study sequences, patterns, and biographies. These include the development of wear analysis; identification of waste within the scheme of production; refitting; experimentation as a way of investigating the relationship between sequences, human action, and intentionality; and spatial analysis to identify consumption, use, and discard (Martinón-Torres 2002; Bar-Yosef & Van Peer 2009). The original scope of *chaîne opératoire* has also been enlarged and refined in a number of important ways that are related to practice. For archaeology, such thinking has influenced trends such as osteobiography (Saul & Saul 1989; Robb 2002: 160) and tools for connecting 'artefacts and people through the physicality of skilled movement' (Walls 2016). Important developments have also been made in allied studies of alimentation (Twiss 2007). As Miracle suggests, 'the study of food would profit from use of the *chaîne opératoire* since food is material culture created by technical and social acts' (2002: 67).

These developments in *chaîne opératoire* have generated a range of conceptual themes and practical approaches that are directly relevant to understanding what butchery 'is' and interpreting the biography of the mark in an archaeological context (Fig. 3.2). The fact that a sequence provides only a partial retelling of the technical activity involved in production (Chazan 2009: 470) has led to a blossoming of endeavour focused on practice, practitioners, and 'situated cognition' (Walls 2016). Within this milieu, not only is the act of production examined more closely, but so too is the variability of learning and enskilment (Pálsson 1994). As such, the context of knowledge acquisition and expression – enactment, a form of embodied action (Varela et al. 1991) – is conceptualized around motion and the interrelationship between the body, materials, social ties, and environment (Walls 2016).

Skill, as a separate entity to enskilment, has also gained attention as a way to understand the feedback mechanisms between the body and the object. By assessing the topic in its own right, it is evident that 'skill is behaviourally developed rather than simply learned like a series of facts, a repertoire of

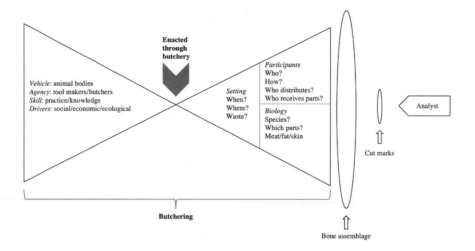

FIG. 3.2 'The biography of the cut mark', showing the relationship between the analyst, the creation of the archaeological assemblage, and the lenses offered by the bone and butchery records.

techniques, or a list of formulae ... As such, skill draws on cognitive *and* motor activities' (Bleed 2008: 156–7). How individuals acquire skills, and the conceptualization of this endeavour around movement, has been instrumental in developing the notion of 'embodied cognition'. If human performance can be compartmentalized into sequences, those sequences in turn can be segmented into 'component actions, gestures and postures that unfold dynamically in space and time' (Marchand 2010: S100). Thus, if it is possible to bring multivocality to the interpretive process by 'reading the archaeological record as text' (Hodder 1982; Tilley 1994), so too does physical practice have the power to communicate (Marchand 2010) through materialisation: *thinking through activity*. These approaches tackle a gap in archaeological research, that is, the absence of practitioners (see Chapter 6).

This is perhaps the most important development to have grown out of *chaîne opératoire*. The benefit of properly situating the practitioner is exemplified by the growth of studies on apprenticeship (Jørgensen 2012; Miller 2012); the connections made between *chaîne opératoire*, apprenticeship, and social boundaries (Arnold 1975; Wallaert 2012); and, perhaps most significant, the acknowledgement that a false dichotomy exists between theory and practice, recognised by academics as well as craftspeople (Jørgensen 2012).

The aim of cut mark analysis has always been to understand why people undertook the butchering we observe in the manner they did, and how this changed through time. This principle is a more effective starting point than to view these traces as descriptive markers in the process of becoming part of the archaeological record (see taphonomic and cognitive approaches in Chapter 2). This also restores how we conceptualise butchery within a sequential frame-work, connecting the activity more closely with processes linked to the time that the animal was alive – even though we are obviously dealing with its death. This aligns with a growing trend in zooarchaeology to emphasize livestock, rather than dead stock (Sykes 2014: 1–2). Particularly when discussing assemblages of domesticates, the overwhelming majority of fauna we encounter have been raised for eventual consumption, and butchery is an essential component of that trajectory. This shift recognizes the incredibly important phase where the animals' body parts are serving myriad nutritional, social, and economic roles. Cut marks then become a distinct biographical *trait*, one of several 'complex markers' that include pathologies and indicators of working activity, or evidence of cooking. By studying cut marks as process (Chapters 5 and 9), we connect to the sequential stages and *biography* of the life, death, and subsequent uses of animal bodies (Saul & Saul 1989; Walls 2016). The sequential nature of carcass processing, constrained and guided in the first instance by the animal's morphology, means that *chaîne opératoire* is an ideal stance from which to study the butchery record (e.g. Seetah 2008; Barton et al. 2013). Where studies of butchery are concerned, *chaîne opératoire* has generally been used in the context of prehistoric research (Miracle 2002: 65–88), although Lignereux and Peters do describe butchery patterns from Roman Gaul as having followed '*les schémas opérationnels*' (1996: 45).

In addition to sequence and pattern, cut marks are a useful and usable dataset for studying gestures and actions, the performance branch of *chaîne opératoire*, situated within a well-established framework: the operationalisation of *habitus*, how one walks, talks, and dresses, 'society written into the body' (Bourdieu 1990: 63). Such thinking has a long precedent, stretching back at least to Aristotle. More recently, Mauss has championed the idea under the guise of 'techniques of the body': 'By this expression I mean the ways in which from society to society men know how to use their bodies' (Mauss 1973: 70). As Mauss goes on to explain, whether discussing swimming, walking, or digging, practice (technique) is dynamic and varies cross-culturally and historically. There is an element of learned behaviour – which one might consider as an apprenticeship period – coupled with a strong connection to societal norms. Thus, the actions of the body are psychological, physiological, and social in their nature (Mauss 1973). Mauss's ideas resonate strongly with a view of craft and practice as guided by the physicality of human actions. However, two further points enhance the applicability of Mauss's approach to the study of crafts like butchery. First, Mauss and other authors recognise the importance of ideology, framed as 'magico-religious' influence (Mauss 1973: 75), as an important feature of craft production. This point is reinforced by ethnographic studies that describe the ways in which belief can dictate tasks and sequences of making (see Chapter 4). Second, and a point not well developed in Mauss's original thesis, is the implicit materiality of gestures (Matthews 2005). The techniques of the body have a well-defined tool-kit, be they spades and shoes or knives and butchers' blocks.

Unfortunately, describing a sequence of carcass disarticulation is usually the extent to which *chaîne opératoire* influences contemporary butchery studies. This effectively illustrates the difference between the spirit and ideology of *chaîne opératoire* and the actual application. *Chaîne opératoire* has been critiqued because the approach is commonly utilised to investigate production, and therefore has a tendency, in practice, to isolate the object from social, economic, design, and technological spheres of engagement (Skibo & Schiffer 2008: 10, 20–1; Hodder 2012: 53). For butchery, this is precisely the type of pitfall that needs to be avoided in order to correctly situate this activity within society.

Thus, where *chaîne opératoire* is concerned, we are some way from including the type of theoretical breadth indicated above into studies of cut marks. The developments in *chaîne opératoire* that could help to situate practice, sequence, pattern, and the butcher have been largely overlooked, with relatively little attention paid to the potential for building interpretation from the 'technological chain' of animal processing.

APPLYING THEORY TO ASSESS BUTCHERY PRACTICE

The theoretical models outlined above, and in particular the *chaîne opératoire* approach as a guiding principle for assessing sequential production, offer numerous vantage points for tackling the complexity of the butchery record. They also provide tools that could lead to a more effective assessment of cut marks (Fig. 3.2). The principles of ethnoarchaeology, i.e., the archaeologist undertaking the ethnographic work themselves rather than filtering interpretation through an ethnographer (Binford 1978), has influenced much research, including my own (from early work by Yellen 1977; see David & Kramer 2001 for a range of examples; Seetah in prep b). Experimental replications focused on a suite of complementary tasks, and the use of analogy, provide an important basis from which to assess the material record. However, methods such as the utility index exemplify the 'cost–benefit' analytical framework that underpins middle-range theory (Trigger 1995). Decisions about consumption and subsequent discard involve a host of

complex economic and social factors, dependent on technological innovation and the place of animals in a given society, in other words, the
ideology driving practice. The context of knowledge is absent; there is no
room to integrate practical know-how, or skill, or to accommodate social
drivers. Trigger saw an expanded view of middle-range theory as the way
forward, one that included the strengths of both processual and post-
processual approaches and situated human behaviour as guided by belief, as
well as social and ecological contexts (Trigger 1995). This stance has led to
insightful interpretations emphasising the ideological and symbolic factors
that govern the production, use, and discard of metals (Lahiri 1995), pottery
(Arthur 2014a), and lithic objects (Weedman 2005).

Another point of incongruence between the Binfordian view of middle-
range theory and an updated and expanded version, in part built from my own
research, revolves around the axiom of static-to-dynamic. Binford saw the
archaeological component as static, and the modern as dynamic. This is a vastly
oversimplified version of Binford's conceptualisation, but the point is relatively
self-evident. Viewed from the vantage point of craft activity, the dynamic
contexts are located at either end of the archaeological record (Fig. 3.2).
Indeed, static in this context is relative; space and time influence this phase.
In fact, witnessed through taphonomy, the static component is also subject to a
degree of energetic activity. This is an important consideration as it is essential
to our ability to apply analogy.

If the strength of middle-range theory rests in the tools it provides for
reconstructing human behaviour, the weakness is the translation of the
material record into untestable reconstructions of that behaviour. Similarly,
in addition to the fact that we have yet to capitalise on developments in
chaîne opératoire, the approach itself is one that is grounded in the production,
use, and discard of tools. Resolving these issues in order to advance our
studies of butchery rests on a deeper and broader assessment of the ecofacts
themselves, situating the practice that cut marks represent into the wider
societal setting.

I have implemented a combined approach that has taken cues from the
approaches outlined above, and in particular *chaîne opératoire*, to capture and
accommodate the complexity of the skilled craft of butchery. Adopting an
expanded view of middle-range theory (Trigger 1995), the empirical strengths
of the approach have been combined with post-processual perspectives
gathered through a suite of ethnographic studies that emphasise the role of
materials in social organization (Trigger 1995) as well as the varied context of
why and how people utilise animals (Crabtree et al. 1989; Hodder 1990:
262–4; O'Day et al. 2004; Marciniak 2005; Arbuckle & McCarty 2014).
The methodological approach rests on the integration of a range of datasets
(Domìnguez-Rodrigo 2002), with experimental studies focused on replicating

the *process* of butchery, not specific marks – as outlined in Chapter 2. This phase also incorporates complementary research focused on the tasks to reproduce metal knives (Seetah in prep c).

It is worth considering the difficulties of assessing activity, given, as it is, a cultural process (Hodder 1992): human behaviour is subjective, but from a technological perspective, the residues need not be. The problem lies in the fact that no uniform links exist between behaviour and artefacts. If the relationship between behaviour and things cannot be modulated to be mean-ingful, we have little hope of accurate interpretation. All meaning becomes essentially phenomenological.

However, practice is clearly cultural within the context of performance and production. It can also serve as a proxy for culture, by providing a measure of activity through basic counts and typologising of cut marks, which can then be assessed statistically to reveal a pattern of butchery, for example. Thus, practice as a feature of behaviour, and therefore culture, is subject to a uniform relationship between action and outcome.

For butchery, employing the same hand, tool, force, bone positioning, etc., will result in the same outcome: marks will be similar *if* we control *all* the variables. This is an impossible scenario to create in the real world. The constellation of factors cannot be constrained nor recreated, and therefore, the cut cannot be replicated in absolute terms. We do potentially have a way of dealing with this limitation. By using an abundance of test cases, derived from the copious faunal remains with cut marks, we can calibrate the differences between our contemporary interpretations and the actions of ancient craftspeople. In this scenario, examples of butchered bone with the process of cut mark creation well documented, defined, and explained serve as individual case studies that allow us to modulate the differences between archaeological and modern specimens. Generalisation is then possible and can be accurate. It is for this reason that replicating the 'process' of butchery (as opposed to single marks) as a means of achieving empirical verification is more effective, leading to accurate approximations of past practice. In this case, a priori input is essential. As archaeologists, we have the opportunity to replicate only a relatively small suite of cut marks, processes, or tech-niques. The professional craftsperson has repeated the same task to the extent that reading the practice from the outcome – reverse engineering – is second nature. Indeed, the knowledge can become so deeply embedded as 'activity' and motion that extracting the underlying thought process becomes difficult.

Thus, equifinality, a major stumbling block to accurately describing butch-ery activity and distinguishing between discrete stages, can be modulated through the integration of contemporary knowledge. Analogy also validates the fact that dynamic contexts exist at either end of the archaeological record, connected through corresponding actions, if not identical cognitive processes. The task of the archaeologist is to bridge the 'dynamic and the static' (Binford 1978; Hodder 1992).

Recasting an axiom from *chaîne opératoire*, the recognition that the knowledge of tool production rests on a three-dimensional concept of mass (Chazan 2009: 469), serves as an ideal starting point to explain that the butcher is negotiating with a three-dimensional object, the live animal and/or dead carcass. Further extending this point, the butcher – as I will outline in Chapters 4–6 – also deals with a range of chronological, social, economic, and ecological dimensions. Thus, we arrive at a three-dimensional object being manipulated in space and over time; time in this context extends beyond the immediacy of the butchery episode itself. Methodologically, *chaîne opératoire* contributes to understanding skill. We need clearer and more robust statements about skill, and enskilment, as this is a tacit aspect of butchery and essential to understanding practice. The didactic process, mediated through apprenticeship, forms the basis for assessing situated and embodied cognition. Coming back to experimental studies, those outlined in this book centre on replicating process, which represents the seminal principle from which *chaîne opératoire* flourished: the idea of sequence. *Chaîne opératoire* brings our attention to the entire sequence and the pattern, rather than a myopic focus on the cut mark. By integrating the context of movement – the 'techniques of butchery' (to paraphrase Mauss) – to investigate use and practice, we lean toward a greater appreciation for the physicality that drives practice. For butchery, functionality is crucial because we are dealing directly with activity and behaviour, not an 'object' per se (Chapter 9). Finally, for historic periods in particular, the individuation of tasks into specialisations recalls to mind the concept of taskscapes. Viewed from this perspective, we can appreciate how the fracturing of 'butchery' into slaughtering, skinning, horn removal, etc., tasks that became the specific role of an individual, forms part of the transition from craft to trade to industry. Taskscapes also describe the relationships between different constellations of activity: those of the farmer, butcher, bone worker, tanner, etc., and how they remain essentially interdependent.

With a combined approach, the model presented in this book bridges scales in data (Trigger 1995), offering a material reflection of skills (Bleed 2008) that is conspicuously absent from existing studies of butchery. The use of analogy covers both economic and social features of butchering and is informed from a uniquely appropriate perspective: practice as learnt through an in-depth period of apprenticeship (Marchand 2010) and applied skill (Walls 2016). By generating examples that connect incidences of cut marks with clear reasoning as to their production, this form of analogy offers perhaps our most important tool for dealing with equifinality. Finally, the methods I propose encourage the analyst to build inference directly from the material during recording, since 'the extraction of technological information from the archaeological data by the *chaîne opératoire* concept relies . . . on an inferential procedure' (Sellet 1993: 110).

CONCLUSIONS

Archaeology embraces the totality of the human past, and we are increasingly responsive to the fact that this includes the activity to produce objects as well as the objects themselves (Schlanger 1994: 143; Martinón-Torres 2002: 35). Through the use of methods such as ethnoarchaeology and analogy, middle-range theory provided an indirect assessment of activity, in the sense that practice was understood as a means to an end, rather than as the aim. In contrast, the principles of *connaissance* and *savior faire* positioned knowledge and skill at the heart of *chaîne opératoire*, which has led to closer attention on the role of practice during production. *Chaîne opératoire* offers one of the most promising directions for the future of archaeological thought and practice, precisely because it is intent on situating craftspeople and their experience through activity-based approaches rather than phenomenology.

Chaîne opératoire operates in the space between material and cultural, providing ways to understand how transforming one into the other is mediated through technology and activity (Martinón-Torres 2002: 34). This translation is a particularly important issue when 'material' is also 'sentient' (Orton 2012), a notion at the core of the question: 'what is butchery'? Technically, the material of the butcher is not sentient for the majority of the time he or she is butchering. Nevertheless, slaughter extinguishes life and, therefore, sentience. This represents a remarkably complex ontological situation that has guided ritual and religious practice and represents a uniquely important human–animal interaction revolving around the power of one species over others. Slaughtering technique provides the practical evidence to address an important question: why do individuals, groups, cultures, and nations kill animals in the way they do?

The fact that some have questioned whether butchery can be considered a taphonomy (Speth 1991: 37) signals a deep-seated concern that we have yet to tackle. If taphonomy led us to a specific view of cut marks (Lyman 1987), then *chaîne opératoire* takes us in new directions. The emphasis on the breadth and complexity of sequence and pattern, biography, and the skill and knowledge of craftspeople serves as a more appropriate starting point to think about butchery as social and economic practice.

A conceptualisation of butchery built from the principles of practice provides the basis to challenge our existing recording systems and fundamentally reinvigorate the subject. We need 'an interpretative methodology and analytical methods capable of forging robust inferential links between the material patterning of technical acts and the *sociopolitical relations of production accounting for them* '

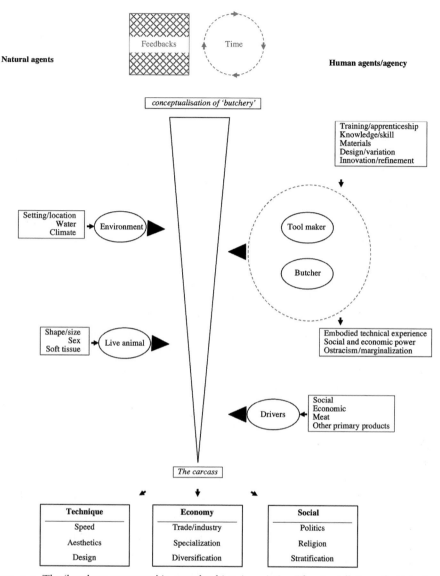

FIG. 3.3 The 'butchery ecosystem' in complex historic societies. The constellation of activity is represented as coupled systems that include natural and human agents, with feedback mechanism crisscrossing between agents in a circular fashion through time and across space.

(Dobres 1999: 124; emphasis added). Dobres's insightful statement describes the conceptual change I advocate. We need to situate the cut mark, the basic unit of analysis, within three levels of inference (Figs. 3.2 and 3.3): (1) locational and typological, (2) interpretation of the functional, technological, and sequential features of activity, and (3) the didactic, sociopolitical, and economic drivers at the societal level.

In essence, these levels capture the principle of why butchery is of significance to archaeological questions revolving around human–animal-environmental relationships. It provides a lens on both social and economic aspects of how humans interact with each other, as mediated through the animals that surround them. However, in the same way that smelting copper requires specific knowledge about ore (Trigger 1995), and clay must be prepared before it can be turned into pots (Hodder 2012: 52), so too must the butcher understand the details of animals and their bodies, as well as the tools used to process the carcass. We need to integrate this cognition in order to accurately situate butchery as the practice-based activity that creates 'meat'. Studying meat, but circumventing the details of butchery, misses an incredibly important aspect of this seminal relationship.

For a newly conceptualised interpretation of the subject, we need to develop an ecosystem approach, integrating the technological, cognitive, and societal milieu within which butchery and butchering are situated. Figure 3.3 describes the butchery ecosystem for a historic context: tools are fabricated, and animals are raised or hunted, which together provide the basic requirements that will catalyse and qualify the practice to follow. A range of drivers guides the hand of the butcher; this cognition is expressed as activity and motion. Space is created and defined in order to accommodate the social and economic relationships between those who make and those who consume. Thus, butchers might be located close to water (Goldberg 1992: 64–6), blacksmiths on the outskirts of towns (Torbert 1985), and farmers of some species in the hinterland, while husbandry of other livestock might be organized around the homestead, even in an urbanized context (Grant 2004: 377). These locational parameters reinforce the notion that the 'landscapes of action and consciousness' are inseparable (Trigger 1995) and become exemplified as 'geographies of slaughter' (Day 2005). Feedback mechanisms include the relationship between blacksmiths and butchers in creating knives, between farmers and consumers in shaping and reshaping the animal, and between the animal's constitution and the tool and the techniques of butchery in modifying the dismemberment process to accommodate changing attitudes to how we assemble, prepare, and consume. This contextual nuance and richness is revealed through the vehicle of the animal's body, expressing the ideology of meat consumption in a given society: flesh is presented as meat, having undergone an essential transformation (Vialles 1994: 5). As a craft, butchery is motivated and propelled dynamically by a concept in the mind of human agents, one that illustrates the deep context of intentionality. Ultimately, practice becomes embedded in society.

CHAPTER FOUR

CRAFT, PRACTICE, AND SOCIAL BOUNDARIES

Except in the processes of butchering, outcastes could not handle foods to be eaten by ordinary people.

(Price 1966: 9)

INTRODUCTION

Both long-standing and more recent engagements with zooarchaeology often pivot between two distinct polarities: the economic and the social (Marciniak 2005; Russell 2012; Arbuckle & McCarty 2014; Sykes 2014). Not surprisingly, the 'social turn' has often focused on topics of shared historic interest, namely, the consumption of animals and their role in ritual and religious expression (Russell 2012: 91; Sykes 2014: 114–32). Meat has long been an important topic (e.g., 'Meat beyond diet', in Russell 2012: 358–94) and builds on a wider and more established body of anthropological literature. Excellent scholarship on the role of food and meat as social capital exists at least since the work of Spencer (1898: religion and military influence) and Veblen (1910: conspicuous consumption). Anthropological enquiry of the 1960s to 1980s further reinforced alimentation as central to human thought and practice, with important theoretical developments made by Lévi-Strauss (1966, 1970, 1978), Douglas (1970, 1975), Barthes (1975), Goody (1982), and Harris (1985). These authors emphasise the ways in which cuisine, dining, and social stratification can be viewed through the medium of food. The role of meat protein in human evolution (Aiello & Wheeler 1995; Stanford 1999; Navarrete et al. 2011), alongside issues surrounding the ethics of consuming animals (Marcus 2005) and the engendered nature of meat representation (Adams 1990; Ruby & Heine 2011), illustrate the diverse relationships that we have with animal bodies and their flesh. Among these varied perspectives, Fiddes's *Meat: A Natural Symbol* (1991) stands out as a seminal text, arguing for recognition of the role meat has played in a range of social settings.

Mengzi wrote: 'a taste for meat is at a world level the most commonly shared penchant' (Sabban 1993: 81). Meat consumption is arguably the most highly charged, widely debated, and heavily moralised food topic, occupying a seminal place within studies of food. However, the trend to emphasize the outcome – flesh, or meat – has tended to draw attention away from how we arrive at the commodity. As I illustrate in this chapter, *butchery* is also 'a highly charged and heavily moralized food topic'.

Anthropological research focused specifically on butchery and butchering takes wide-ranging forms. For example, ethnographic studies have explored the relationship between the butcher, the animal, and the location of carcass processing (Mooketsi 2001), as a way of connecting butchery to wider social drivers. The varied interconnections between animal processing, the animal's flesh, and the construction of taboos have also been an area of keen interest for ethnographers (Ross et al. 1978; Forth 2007; Luzar et al. 2012). This broad category, not necessarily anthropological in methods but in research focus, also includes contemporary studies of the history of the industry (Watts 2006; Lee 2008; Fairlie 2010; Pacyga 2015) and those who work in it (Fogel 1970; Pilcher 2006). Of the latter category, perhaps the most influential work of the time was a fictitious narrative called *The Jungle*, by Upton Sinclair (serialised in 1905, published in 1906), which highlighted the working conditions of the then modern industry and led to sweeping reforms in the United States.

I have deliberately brought together a wide range of perspectives to emphasise the diversity of relationships that exist and revolve around butchery. In all the above cases, whether directly or indirectly, butchering is central. The above represents a top-down interpretation of carcass processing, considered from the perspective of the place of meat in society. The complementary viewpoint from archaeological remains will not be as clear. It is easier to see how cut marks can be indicative of specific sharing practices in non-industrialised cultures, but perhaps harder to note how butchery processes may provide a lens on industrialisation itself or those involved in the trade. Such ideas offer testable hypotheses and raise the bar in terms of the type of questions archaeologists investigate. As Russell (2012: 3) suggests, moving beyond 'protein and calories' allows us to view a more complete repertoire of factors that influence how we interact with non-human animals.

Thus, in contrast to the social equality that actor–network theory (ANT) presupposes, an important aspect of butchery revolves around the power and influence of social attitudes toward aspects of the craft and those involved in it. Meat is featured within a range of settings, from civic identity (Billington 1990; Seetah 2010) to negativity associated with working closely with animal bodies, their waste, and death (Charsley 1996). It serves as a dominant means of social differentiation (Groemer 2001). However, in none of the examples described so far are we actually relying on cut marks to deepen our understanding of the role of meat. An excellent illustration of this point is Mooketsi's (2001) research in Botswana, whose analysis evaluates the role of butchery and the butcher, but includes no assessment of cut marks.

Focusing on non-mechanised settings and emphasising 'social practice', the following explores how cross-cultural perceptions of 'self' and 'other' are codified and expressed on the basis of interactions with animal bodies. Applying the principles of analogy, ethnographic enquiry, and ethnoarchaeology, the discussion positions butchery within the larger socioeconomic context, spotlighting an aspect that is central to how the craft developed: the relationship between those who work as butchers and the society in which they live. This offers zooarchaeologists a way to connect butchery as 'sociotechnological practice' to society and to situate the practitioners involved in various crafts. This stance also provides a bridge between social and economic sides of zooarchaeology and resonates with developments in *chaîne opératoire* that are aimed at uniting practice and the practitioner.

This chapter engages with ideas drawn from social studies. In this regard, 'status' relates to a socially defined position in society, whether small scale or large; 'social differentiation' describes the process by which different statuses develop in any given society; and 'social stratification' refers to relatively fixed, hierarchical arrangements in society by which groups have varying access to social and economic capital. In addition, I also borrow from social studies to distinguish between caste and class systems, the former denoting that an individual's standing in a stratified system is ascribed, while the latter emphasises that a person's place within their stratification can change according to their achievements (Saunders 2006: 1–27).

CRAFT, MARGINALISED GROUPS, AND ARCHAEOLOGY

The study of humans and animals in the past could benefit from adopting tried and tested approaches developed for other material sciences in archaeology. Studies of metal (Lahiri 1995; Derevenski 2000), ceramics (Jones 2001; Arthur 2014a), and lithic implements (Weedman 2006) have all shown that ideology and cosmology are critical drivers of production, assembly, use, and discard of objects: 'The ideological underpinnings of metalcraft and the perception of metalcraftpersons provide the context and justification for understanding some of their practices' (Lahiri 1995: 116). In numerous cases, ethnoarchaeology has helped to reveal ideology in ways that can then be directly applied to the archaeological context (Arthur et al. 2009). Weedman, for example, points to how an ideology of purity drives segregation of certain artisanal groups who form part of Gamo society, in Ethiopia (2006). A number of features seem to illustrate salience across geographic, chronological, and cultural boundaries, which may have genuine significance for understanding 'craft' in a more complete way. Indeed, approached in this way, situating craft becomes an essential feature of human evolution, expression of social norms, and economic development. Studies have shown how relationships within discrete 'craft groups' influence the transmission of practice. In India, for example, enskilment is expressed along caste or caste-like lines; 'occupational speciality is learned and shared within endogamous groups which often have distinctive names and sometimes live in circumscribed areas' (David & Kramer 2001: 308). This is equally the case for relationships between craft groups. In cases where ideology drives segregation of certain crafts – seen as distasteful, unclean, or impure – which are then collected into one location on the periphery of a settlement, this can lead to between-group marriage, with attendant ramifications for learning, acquiring, or refining skills (Sterner & David 1991).

Zooarchaeologists recognise the importance of ideology as a driving force in the creation of bone assemblages. However, these have generally been restricted to specific cases, particularly as they relate to religious and ritual practice (Greenfield & Bouchnik 2011: 106–10). The potential benefits of adopting the types of detailed ethnoarchaeological studies that have gained attention in the study of pottery and lithics are impressive, particularly in situating the human half of the human–animal relationship.

Carcass to Meat – A Hard Act to Swallow?

The social and ethnographic contexts surrounding the practice of butchery reinforce the notion that craft is embedded within cultural milieus. In the discussion that follows, the lens falls firmly on the human agent, rather than animal bone or cut marks. For some societies, the corporeal animal carcass is seen to retain qualities that are considered base, attributes that the spiritual human should avoid. The examples are drawn from in-depth contemporary case studies that discuss marginalised groups from a range of culture contexts. Drawing on and generalising from this extensive literature, I examine how relations with meat, slaughter, animal bodies, and body-part processing have formed the basis for negative attitudes toward those individuals involved in these crafts. In the interests of further expanding the picture, I also draw on cases where those involved in animal slaughter and processing derive from the religious elite or economically privileged portions of society. A brief survey of post-medieval to contemporary examples drawn from Europe, in particular Britain, and the United States is presented to contextualise the ethnographic and historic case studies with modern industrialised examples. These cases illustrate some of the ways that the craft of butchery changed, once systems were in place to deal with those aspects of butchery that caused greatest nuisance. This final category is grounded more in the development of capital and less on belief, although religion and ideas of sanitation and cleanliness were still important drivers.

Although this is not a universal, it is often the case that societal attitudes toward those involved in animal processing can be vehemently negative. Religious principles and moralised understandings of cleanliness serve to create and promulgate identities that define individuals, and even ethnicities, in terms of their associations with butchery, within larger contexts like regions or nations. Positive attitudes can be equally socially relevant in some settings and likely rest on the fact that the butchers were valued because of their skill and knowledge, as well as for the economic capital that their enterprise made possible. In some instances, this also extended to increased social value (Billington 1990). I focus more of my attention on examining and understanding negative attitudes, as these have harmful outcomes for the people involved. Further, the consequences of these negative attitudes have tended to concretise practical tasks and material objects within and around the marginalised groups' environment, particularly through endogamy, and this has particular relevance for archaeologists.

A number of characteristics appear to drive negative attitudes to carcass processors. For many, meat consumption is associated with vitality and good health (Brantz 2005). In contrast, vegetarianism, veganism, and modern concerns about the implications of excess meat on human and environmental health – to name a few examples – all point to the ways in which individuals and societies wrestle with the act of consuming flesh (Rozin et al. 1997; Ruby & Heine 2011; De Backer & Hudders 2015). Whether viewed from the perspective of individual or societal unease, one facet of consumption has historically been of particular concern: slaughter. It has been suggested that meat consumption decreases as societies advance technologically, while sensitivities toward the act of slaughter itself increase (Ellen 1994: 203). This is far from universal. Indeed, certain cultures consider the suffering of the animal to be an essential aspect of ritual sacrifice preceding consumption. For the Masa Guisey of Cameroon, for instance, the offering of a chicken involves a slow death, the bird plucked and singed while alive (de Garine 2004).

To take another example, for many practicing Buddhists and Hindus, it is precisely this suffering that underpins their cultural attitude against the taking of life for consumption. Similarly, for Christians, Jews, and Muslims, religion plays an important role in shaping attitudes toward meat eating or those who undertook slaughter. However, in each case important nuances exist, particularly in accordance with interpretations of religious texts (Bouchnik 2016: 306). For Judaism and Islam, the act of slaughter was an integral aspect of religious practice (Greenfield & Bouchnik 2011: 107). In contrast, during the Middle Ages in Europe, as part of a more general emphasis on abstinence and material paucity to avoid the sin of gluttony, certain orders of monks refrained from eating meat. The situation is further complicated due to religious ideology. In medieval Latin Christendom, meat could be associated with moral pollution as it was considered to promote sexual appetite. However, at the same time, many religious groups, particularly high-ranking clerics and abbots, regularly consumed meat (Ervynck 2004: 216–21). The view of animals also needs to be situated: fauna walked on all four feet and were closer to the earth, and therefore base. Worse still were snakes, which crawl and were the 'adulterer' (Crane 2013: 100). Indeed, by the Carolingian period, the rule of Saint Benedict was widely adopted in Western Europe, resulting in the prohibition of meat from quadrupeds for monks in good health (Schmitz 1945). Ultimately, this may help explain the subsequent dependence on fish (Ervynck 2004: 219).

Faith systems from Asia illustrate a number of examples of 'sins by association', discussed in the case studies presented below. Becoming morally polluted, as per the Christian example from the Middle Ages, is reinforced by a range of Asian examples that emphasise physical pollution, as a consequence of working with animal bodies and their blood and waste. This illustrates the complexity of the situation. In numerous cases it is not meat that is the problem, nor its consumption, but rather the specific interaction that the human has with the beast: killing. Slaughter, and preparing a 'corpse' through evisceration, exsanguination, and skinning, incorporates a different set of acts than preparing 'meat' for consumption. More recently, these 'wet' acts, in societies where butchery and carcass preparation have become institutionalised (Lee 2008: 4), have been removed from public view. The flesh is transformed into meat behind closed doors (Vialles 1994: 5). The preparation of meat for consumption is clearly an act of butchery, but one associated with food: through cuisine, meat is prepared to be eaten.

Economy also plays an intriguing role. It is revealing that in many cases considerable wealth was involved. Monastic communities invested large amounts of capital into fish, fowl, and game, all examples of flesh that they could consume (Murray et al. 2004). Butcher castes in Asian and African contexts monopolised the processing of animals and could become very wealthy. However, in these latter cases, the financial capital could not erase the perceived notions of defilement nor permit those involved in the profession to transition to a better social class.

Finally, while the following accounts are drawn from contemporary or recent historic accounts, we should be alerted to the likelihood that this type of social stratification might reveal itself to have been far more prevalent in the archaeological past. Two accounts from the Canary Islands, recorded during early contact between pre-Hispanic peoples and Castilians, provide tantalising food for thought for historical zooarchaeologists:

> The job of butcher was perceived as vile and crude, and was performed by the lowest man; and he was so disgusting that he was not allowed to touch anything, to him was handed a wood stick to point at the things he wanted. He was not admissible inside the houses, or to mingle with people, but only with other colleagues of his profession; and as a reward for his submission, he was given the things he needed. And none was allowed to kill any animal, unless in the places arranged by the butcher, and in these houses youths, women and children were not admissible. (Historia de la conquista de las siete islas de Canaria 1602; Cioranescu 1977)
>
> [T]he worst damage they could enact on the Spaniards captured by them (the indigenous peoples), was to scalp them and make them slaughter animals for meat, and to cook it, and to roast it. (Canarias, Crónicas de su conquista 1500/1525; Padrón 1993)

These fascinating reports speak of 'untouchability' within a wholly unrelated context to that seen in India or other modern analogues (Price 1966: 39; Amos 2011: 29–32). They also exemplify a phenomenon that is repeated in many of the cases that I will present: carcass processing is defined culturally as among the basest and most menial tasks that an individual can perform. Finally, another remarkable similarity is that a deeply ingrained hierarchy seems to be a common feature of the societies outlined, which was the case for the agricultural, pre-Hispanic peoples of the Canary Islands (Morales et al. 2009).

CASE STUDIES IN SOCIAL DIFFERENTIATION

Of the three monotheistic religions, none would prohibit a butcher from entering a place of worship. Indeed, for Islam, Judaism, and Christianity, a case could be made for the fact that meat processors and butchers enjoyed a relatively well-regarded status, whether due to economic factors or because they performed a task that had its roots in religious text. However, in India, Japan, and Korea, for example, religious and societal ideals created a complex system of separation for those whose trade resulted in an association with animals and animal bodies. In particular, butchering of animals appears as an almost universal mechanism used to designate outcaste status (de Vos & Wagatsuma 1966: 7). In all these cases, a social hierarchy based on heredity, in combination with a religious ideology based on a tenet of cyclical rebirth, led to the establishment of systems of untouchables and outcastes that extend to the present day (Groemer 2001; McClain 2002: 100). The following survey commences with India, partly as it has the largest and oldest 'outcaste' population of any contemporary society, as well as one of the most complex (Mendelsohn & Vicziany 1998: 1–6; Jodhka 2012: 1–33; see also Appadurai 1988). In India, the caste system and the plight of the untouchables has been one that has received a great deal of public attention, particularly since the time of Gandhi (Adams 2010; Das 2010).

India

The relationship between abstinence from meat eating and Hinduism is well known and complex (Staples 2008) and, as one might imagine, has ramifications for those who are involved in carcass processing. One way to achieve virtue and promote a positive karma for Hindus is to abstain from eating meat. However, equating modern vegetarianism with Buddhist and Hindu traditions is not an accurate comparison (Masri 1989: 49). Neither Buddhists nor Hindus are against the eating of meat per se (although this point is contested, see Page 1999: 122); it is the sanctity of life that prompts a vegetarian diet in these groups. Both faiths permit the eating of meat that has not been killed expressly for the purposes of the individual's consumption (Hopkins 1906; Chakravarti 1979; Masri 1989: 49). Furthermore, within the Hindu caste system, a person of the warrior caste would be considered to be acting within the remits of their nature to eat meat, for example, to improve physical conditioning (Khare 1966).

India's endogamous system of social hierarchy is based on various attributes that include vocation, religious belief, and linguistics. The Jāti (a group of communities) is divided into four main castes to which all individuals belong. Each Jāti is associated with a traditional trade, job, or tribe. Those of the lowest Jāti were invariably involved in professions that were considered menial, such as blacksmiths and weavers (Natrajan 2012: 29–30). Within this social stratification exists a level that is beneath the lowest of castes: the untouchables.

Caste differentiation is socially sanctioned and condoned to the point that those involved in the animal trades are regarded as inhuman (a theme that recurs in both Japan and Korea). In part, this was based on fatalism (Elder 1966). The individual must have been wicked in a previous life; it was their karma. Those involved in animal trades were considered base, untouchable because of their profession and both physically and ritually polluted (Passin 1955, 1957; Mendelsohn & Vicziany 1998: 7). There is a difference between the notions of caste and 'outcaste' (which is a better descriptor for the Japanese and Korean examples) and untouchability. Though untouchables worked in a range of professions, including sweeping and laundering, those involved in

animal slaughter and processing were considered tainted because they performed a task that was defiling (Price 1966: 39). To illustrate this, Brahmins
(the highest caste composed of priests) who killed animals for consumption
were considered inferior to other Brahmins who did not (Dumont 1980: 358).
It is the interaction with animals, and specifically the act of slaughter, which
underpins the idea of moral pollution. Those who repetitively performed such
tasks were not able to remove the taint that killing left on them and they
deserved to be both reviled and ostracised. Although they might attain wealth
and prosperity, relative to their station (Siddiqi 2001), they were never able to
remove the desecration of their actions.

Social differentiation based on caste in India is expressed in a number of
physical ways; in the worst cases these are violent (Shah 2006: 97; Natrajan
2012: 11 and 170n3). Lower castes could be precluded from entering temples;
they could not associate with those of a higher caste nor take meals with them
(Mines 2009: 1). Their social positions become fixed from one generation to
the next due to endogamic marriage. For the untouchables, as the name
implies, no physical contact could take place with members of a higher caste
(Charsley 1996). For those untouchables who dealt with animals – the slaughterers, butchers, tanners, and leatherworkers – the act of killing animals and
working with their body parts was seen as penance. Thus, there was no
compunction in treating them with disdain. Their release from this particular
status would come with death, and, provided they were virtuous in their
present lives, their rebirth would be more favourable.

Japan

As in India, similar cultural and religious ideologies combined to create Japan's
own outcastes, the Eta (meaning 'defilement abundant'). During the Toku-
gawa Period (1603–1868) this group's ostracism effectively became codified in
law and practice (De Vos & Wagatsuma 1966; Groemer 2001; see Amos 2011
for a recent and important work on the Burakumin). The situation in Japan
was different to India in that two separate religious ideologies, analogous but
fundamentally distinct coalesced to influence the creation of social outcastes
(De Vos & Wagatsuma 1966: 3). An underlying tenet of Shintō is avoidance of
impurity (McClain 2002: 100). While Christianity views cleanliness as next to
godliness, in Shintō cleanliness is godliness (Passin 1955). In effect, those who
worked with animals were not in a position to be able to attain purity; they
were irrevocably impure, dirty, and corrupt. With the arrival of Buddhism and
an ethic that all life was sacred (Donoghue 1957; McCormack 2013: 39), those
involved in animal trades, especially those who had to carry out the physical act
of slaughter and the letting of blood, were even more reviled and considered
to be polluted. Eta were composed of a range of professions. Thus,

leatherworkers, tanners, and butchers were grouped with unlicensed prosti-
tutes. This indirectly illustrates the social group that those in the animal trades
were identified with. Another group of outcastes, termed Hinin (non-human),
included professions considered unsavoury, such as monkey trainers and actors.
As with the untouchables of India, the condition was considered hereditary,
but only for the Eta; the Hinin were outcastes because of the professions they
found themselves in due to poverty or other reasons (Passin 1955). In an
interesting parallel with numerous European examples, Eta involved in animal
trades were called on to be executioners (Groemer 2001).

The clear division between the treatment of Eta and Hinin and of 'normal'
people exemplifies the attitude to this outcaste group. Eta were historically
excluded from censuses or, if recorded, were tabulated separately from
'people'. Their villages were not recorded on maps or the maps themselves
were abridged to eliminate the location of their dwellings. These attitudes
were legally as well as socially sanctioned. In 1859 a magistrate was asked to
rule on the case of a non-Eta killing an Eta. He decreed that an Eta was one-
seventh of a normal person, and therefore the non-Eta would have to kill six
more Eta before he could be punished. The descriptive terms Eta and Hinin
are themselves highly offensive, equivalent to racial slurs. This has more
recently resulted in the use of terms such as Burakumin (community people),
which is considered less offensive.[1] The mark of the beast was firmly
reinforced by a variety of derogatory animal associations. Eta would be
counted with the same classifier as used for animals (Hiki: 匹). The common
insult for Eta was to hold four fingers in the air or refer to an Eta as 'four'. This
was intended to reinforce the association between animals and Eta, suggesting
the four legs of an animal, or that the Eta had only four fingers, one less than
normal people (Passin 1955). These attitudes are so pervasive that even in the
more recent past, Eta have considered themselves to be base, identifying
strongly both with their activity and with animals. During interviews con-
ducted with members of the Burakumin, when asked if they are the same as
other, non-Eta, people, the response was that they were not: 'We kill animals.
We are dirty, and some people think we are not human.' When asked why
other people are better, 'They do not kill animals' was the response (Donoghue 1957: 1015).

Korea

The Korean outcaste system, established during the Koryo Period (AD
918–1319) reflected many similarities to those seen in Japan. Like Japan, the
incumbent belief system, in this case Confucianism, upheld ideals of upper and
lower orders and served as a basis for social differentiation. With the arrival of
Buddhism and the doctrine of the sanctity of life, those 'lower orders' of

people involved in animal trades, who could never truly attain virtue according to Confucianism, became outcastes (Passin 1957; Neary 1987). As in Japan, two groups of endogamous outcastes were created: the Paekchŏng (or Baekjeong), who were butchers and analogous with the Eta, and the Chianin (or Jaein), incorporating the same vocational groups as the Hinin.

More so than the Indian and Japanese examples, the Paekchŏng were charged with all aspects of animal processing. They were the leatherworkers, tanners, and of course butchers (Kim 2013: 15). They were also dogcatchers and responsible for the removal of animal carcasses. Of the outcastes mentioned, they were arguably the group most closely associated and defined by the subject of their trade. Again, the theme of moral inadequacy was exemplified in their role as executioners. Although the creation of this outcaste group effectively permitted the Paekchŏng to monopolise certain trades, this did little to compensate for the degree of social ostracism. The measures to delineate these outcastes were extreme. They were expected to show acute servility, bowing to all those from upper classes, including children. Their isolation from society was promoted in a range of forms. They were not permitted to wear silk, nor the top hats that all married persons wore, and could not roof their houses with tile. Thus, the group could be easily identified through their clothes and homesteads. Furthermore, they had to bury their dead in segregated plots and were not permitted to use funeral carts (Passin 1955).

Tibet and China

Tibet shares similarities with India, Japan, and Korea. In Tibet, the subjects of marginalisation are the Ragyappa (or Rogyapa), who are responsible for slaughtering and the removal of animal carcasses, as well as preparing human corpses for sky burial (Gould 1960). However, China has no such social outcastes based on animal trades, although outcastes exist as a consequence of other social attributes (Hansson 1996). The underlying difference is that China has a system of hierarchy based on virtue (Gould 1960) rather than heredity, which can be considered a class, rather than caste, system.

African Examples

African examples of endogamic caste systems, sanctioned and based on occupation (Weedman 2006; Arthur 2008, 2014b), illustrate salience in many regions of the world, reinforcing multiple cases of social causation. Indeed, though castes have been considered a pan-Indian phenomenon, the similarities are remarkable, with all criteria of true caste met save the influence of Hinduism (Pankhurst 1999).[2] Ethiopia has been a particularly noteworthy case (Halpike 1968; Todd 1978).

Among the indigenous Manga people of the Central African Republic are the Hausa, who entered the region following European colonialism, circa 1900. Although the Hausa as a group are not defined as 'butchers', within their population is a subgroup of professional butchers. This subcaste incurs both inter- and intragroup prejudice as a consequence of their association with animals, the act of killing, and the letting of blood. This is despite the fact that their profession allows them to attain considerable wealth. Their base nature is considered adequately strong that they have no obligation to partake in religious ceremony or *hadj*, and the wives of Hausa butchers are not secluded in walled enclosures as are the women of the Manga, which is an expectation under Islamic edict (Horowitz 1974).

The polluting effect of slaughter forms the driver for other butcher castes. In Senegal, the 'griot' serve as butchers among the Wolof and Serer peoples. In northern Cameroon a sacrificer, who formerly would have been a slave, performs the act of slaughter in lieu of the religious chiefs who are prohibited from killing or handling meat (de Garine 2004).

In Ethiopia, the role of the Fuga in Gurage society highlights the issue of hygiene and cleanliness that often underpins cultural attitudes to butchery. The Fuga, a somewhat generic term for indigenous groups in the region, are considered to be primitive hunters that were subjugated during the historic past. In this case, the Fuga were first conquered several centuries ago by the Sidāmo, and subsequently by the Gurage, and now live as part of the Western Gurage Tribes. Craft specialisation typifies their role in Gurage society. In particular, their role as tanners, *Gezhä*, is a key profession, and the skills of this craft are passed on from generation to generation. The Gurage consider the specific roles taken up by the Fuga to be vile, and this expresses itself through a disdain for any physical contact. The fear of contamination is so severe that Fuga can only enter the house of a Gurage if invited, following which the homestead is ritualistically cleansed (Shack 1964). The Fuga are also required to practice endogamy.

PROJECTING THE CHARACTER OF 'THE BUTCHER'

In contrast to the cases outlined above, a brief survey of Judaic, Islamic, medieval, and modern Western perspectives provides counterpoints to the types of negative associations that have been attributed to those involved in carcass processing. In some cases, the same roles and practices are used as a powerful means of self-identification and nationhood. It should be made clear that these examples do not mimic the cases already discussed; the cultural context is wholly different. Rather, the following illustrates the diversity of the lived experience. Here, positive associations are discussed as they relate, once again, to religious ideology but also to developing modernity. It should also be noted that negativity is not absent in these cases. However, in addition to negativity being predicated on society's view of those who butcher, and attendant associations with death and bodily waste, there were also economic drivers, such as the fear that butchers deliberately sold diseased meat for general consumption. The point to note is that these adverse attitudes did not result in the level of social ostracism observed in the African and Asian cases.

In Jewish and Muslim traditions not only must the animal be consecrated prior to slaughter, but also only specified men, often a rabbi in the Jewish faith, are permitted to carry out the act of killing (Masri 1989: 113; Cohn-Sherbok 2006: 86). Slaughter, if performed according to correct practice, is not seen as defiling. In fact, Judaic and Islamic practices demonstrate the polar opposite of the Asian perspective. Here the groups identify themselves by the specific way in which they kill and process animals. The practices associated with keeping kosher are thus used to preserve group identity (Feeley-Harnik 1995; Brumberg-Kraus 1999; Buckser 1999). However, where these religious groups form a minority within other communities, the techniques and practices of slaughter have drawn attention, emphasising the differences between groups and reinforcing the ideology of 'the other'. Attitudes toward shechita in Europe and America have highlighted the association between anti-Semitism and the anti-shechita movement, veiled behind the rhetoric of prevention of animal cruelty (Lewin et al. 1946: 21; Judd 2007). Similarly, in parts of Europe and France in particular, the Muslim *fête du mouton* has received particular attention as a consequence of the overt nature of sacrifice, and slaughter, within urban enclaves (Brisebarre 1998).

Capital growth, modernity, and industrialization also influenced the way those involved in carcass processing were viewed. Using Britain as an example, individuals working in the animal trades and associated with the various guilds were part of powerful institutions. Members of the Butchers' Guild, which attained Royal Charter in 1605 (Jones 1976: 33), enjoyed a large degree of public goodwill and were integral to particular festive occasions (Billington 1990). They were also important to the economy of major urban enclaves (Watts 2006) and provided a service that received the attention of nobility and even royalty (Jones 1976: 181). On a more fundamental level, butchers them-selves were valued, and this can be seen in the fact that their shops were often located on the high street. Indeed, by the 1600s, at a time when many trades were restricted to a single street, which could quite literally be a back alley, butchers in London had established themselves in an internationally recognised market, Smithfield (Forshaw & Bergström 1980: 35). To have had a son apprenticed to a butcher appears to have been an aspiration (Jones 1976: 15). From the later medieval period in Britain and other parts of Europe where similar guilds were established, butchers and others involved in the animal body-part trades could become economically salient and enjoyed a respected status (Yeomans 2008; Seetah 2010).

Negative associations were not absent, and these too are revealing: 'Many have a great aversion to those whose trade it is to take away the lives of the lower species of creature. A butcher is a mere monster, and a fisherman, a filthy wretch,' wrote the Reverend Seccombe (1743). At this time, the function of a butcher did not always involve killing the animal, a task that was often undertaken by a slaughter man (Rixson 2000: 96). The point is that negative attitudes could derive from associations with death, although the evidence suggests that both legislators and the public were more concerned with the quality of the product, pricing, and corruption. Numerous cases exist where butchers were held accountable for selling spoiled meat or flouting laws designed to prevent the sale of meat during Lent or on Sundays (Jones 1976: 123).

Finally, although rooted in processes that have a deeper antiquity, at least since the post-medieval period in Europe and North America, the drive toward modernity has had major implications for the place and role of the contemporary butcher (Otter 2008; Pacyga 2015). The removal of animal slaughter from the high street is strongly linked to legislation designed to control physical waste and pollution (Perren 2008). Vialles suggests that in post-medieval Europe, 'an ellipsis between animal and meat' (Vialles 1994: 5) led to the establishment of designated abattoirs and the categorical removal of slaughter from the urban domain and public sight (Lee 2008: 47). The situation is an intricate one. For Britain, there was a strong connection between waste and disease, in a modern sense. This resulted in reforms to public health legislation, led in part by Edwin Chadwick's report on the sanitary conditions of the working poor (Chadwick 1843). However, the circumstances of change also revolved around the transformation of humanitarian attitudes toward animals and a process to 'civilise slaughter' (Otter 2008: 89; MacLachlan 2008: 107). Also important were the economic drivers that pushed for a greater regimentation of the process of slaughter and butchery (MacLachlan 2008: 107–27). These attitudes, during a period of globalisation of trade and intensification in local urbanization, found ready adoption in many parts of the United States and Mexico (Horowitz 2006: 1–18; Pilcher 2006: 1–15). However, with the industrialisation of meat production, i.e., increased mechanisation of carcass processing, the butcher and the craft were marginalised in an entirely new way. In effect, the skill disappeared. In its place, individuals undertook precise and specific mechanical steps within the process, and an increasing specialisation of tasks and functions occurred, resulting in a mode of butchery that mimicked the factory line (Gerrard 1964: 86–101). 'Once the slaughter began, it seemed endless. In the cattle kill ... the line moved with ferocious speed never before demanded of workers' (Pacyga 2015: 70).

While this survey has marshalled a variety of perspectives that could usefully be employed to better position the butcher, it is also important to recognise limitations. Coming back to the Asian examples, professions that to some minds are *dégoûtant*, such as collecting 'night soil' (human excrement), did not result in outcaste status in Japan (Passin 1955). Conversely, leather workers and tanners were also considered as outcastes even though they would not have killed the animals themselves. Shoemakers were considered base because Confucianism considers the feet as an opposing force to the head and therefore unvirtuous. In these cases, sinning by proxy led to these groups also being identified as the moral 'other' (Page 1999: 128–9). The examples illustrate how facile the reasoning could be and how easy it was for outcaste status to be attributed; such convoluted reasoning would render any archaeological view opaque.

Furthermore, deciphering the view of a particular profession in the past requires a large measure of caution, particularly when relying on historic accounts. Swift wrote that 'physicians ought not to give their judgment of religion, for the same reason that butchers are not admitted to be jurors upon life and death' (Swift 1843: 385). On the surface this may suggest that the very nature of those who killed, butchered, and processed carcasses was called into question and that the morality of butchers was heavily scrutinised as a result of their profession. However, Swift, an early modern satirist and a cleric, sought to provoke a reaction, and more importantly, his comments were probably aimed directly at physicians and not at butchers. With increased secularisation of society during this time, and growing reliance on doctors to administer a 'cure' for ailments, rather than a vicar, it is plausible that these statements are a reaction to the threat to his profession and the sanctity of religion.

STRATIFICATION THROUGH THE LENS OF ARCHAEOLOGY

Inequality, in one form or another, is universal. As archaeologists, how can we approach a more complete understanding of social divisions, and how do animal remains fit into this paradigm? Once again, the emphasis rests on the knowledge and skills of butchery and butchering itself, but viewed from the external perspective, by society at large. In this way, we observe how practice can shed a light on a more complex and universal topic: identity.

An important body of research in zooarchaeology has looked at the individuals who process animals from economic and cultural points of view (Mooketsi 2001; Pluskowski 2007: 110; Yeomans 2007: 98). Building on this body of literature in terms of scale is research that links bone distribution and site organisation to interpret occupational topographies (Benco et al. 2002). By adding to this archaeological view, the preceding examples seek to better understand mind-set – as it relates to those who handle carcasses. This then serves as a point of departure for understanding social organisation. All share three common factors that are relevant within an archaeological context. They exhibit culturally sanctioned attitudes toward hygiene, are driven by religious ideology, and would likely leave specific material signatures discretely positioned within the landscape. Bolstering these points is the fact that the perpetuation of outcaste status is invariably reinforced by endogamy. Thus, the association between activity and social stratification becomes rooted in heredity and fixed within specific locations over generations.

These features provide a usable framework from which we might better understand social structure from cut mark data. The following depends on one assumption: in cases where particular practices are adhered to, perhaps over generations, these would illustrate specificity in the modes of processing that were particular to individual groups. Such 'archaeological butchery groups' could reveal much about the agent as well as the relationship between agents at different social levels and the religious practice of a site's inhabitants. It also potentially situates the place of technological knowledge within the context of expressing social ideals.

In the above case studies, religion and cosmology have been underlying drivers. Judaism and Islam follow strict rules that dictate slaughter; for Hindus and Buddhists, killing itself is sinful. In contrast, although there are guidelines directing which categories of animal flesh can be consumed and when, Catholicism is relatively free of regulation as it pertains to carcass processing. Thus, one would expect to see 'normal' practices for slaughtering, which is precisely what was noted from Tarbat, Portmahomack, in Scotland. The site, perhaps the earliest Pictish monastery and as such an important example of a monastic enclave (Carver 2008), demonstrates how, in the absence of ecclesiastic indoctrination to define methods of slaughter, common practice prevailed. Poleaxing was employed to kill cattle, and pigs were slaughtered by sticking (Fig. 4.1a,b), two methods widely used for slaughter of these species. Contrasting these examples with those where the technique for slaughter is more rigidly defined provides evidence for religious drivers of slaughter (Cope 2004; Greenfield & Bouchnick 2011).

Additional features that might be indicative of social differentiation include divisions of labour and evidence for hierarchical consumption. The scapula from medieval Toruń Town, Poland (Fig. 7.1, Chapter 7) and research drawn from medieval and post-medieval Britain (Yeomans 2007) illustrate differentiation of tools and crafts. This points to a likely compartmentalisation and segregation of animal processing, and potentially supports the involvement of different individuals. This is not in and of itself indicative of social stratification, but rather, in a highly specialised system that has developed over generations, we may see evidence of well-defined tasks. Finally, there is good evidence for

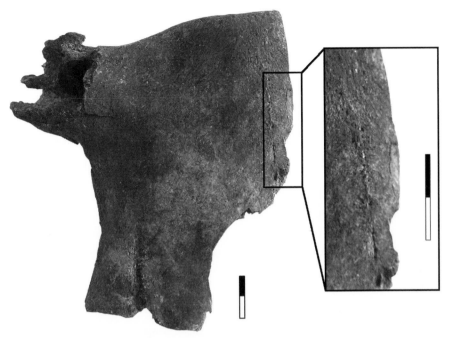

FIG. 4.1a Poleaxe use on cattle (cranium); note the semi-circular puncture shown in the inset, as well as the depressed, but articulated, fractured bone suggesting radiated impact, typical of poleaxe use.

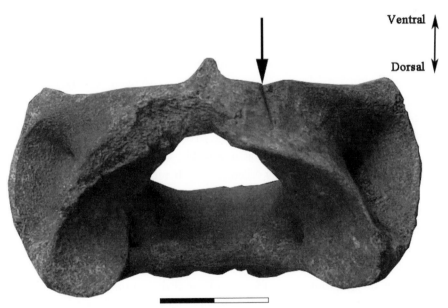

FIG. 4.1b 'Sticking' used to slaughter a pig, ventral aspect of atlas bone (first vertebrae) shown. The mark (arrowed) shows that the point of a blade was inserted into the throat; probably with the animal laid on its back.

differentiation during consumption, as reported by Stokes (2000) from Roman Britain, and Sykes (2007) for the medieval context.

Moving from the cut mark data to spatial context, and in recognition of miasma theory (Rawcliffe 2013: 8–9), one outcome of having to deal with effluent waste from animal processing is that these trades are usually grouped together. Thus, occupational segregation creates vocational topographies, or 'geographies of slaughter' (Day 2005). These have been documented from a wide range of settings, in Europe, i.e. Britain (Goldberg 1992: 64–6 for York; Yeomans 2007: 103 for London), Asia (Passin 1955 for India; McClain 2002: 101 for Japan), and Africa (Horowitz 1974 for Niger; Weedman 2006 for Ethiopia). Spatial and logistical practicalities, such as the need to be close to water, serve as important drivers (see example from York, in Goldberg 1992: 64). However, social factors, for example, the status of those involved in these professions, must also have gone hand in hand with these practical considerations.

Routes to Professionalisation of Practice

Thus, to return to the archaeological case studies, the butchery evidence reveals much about the technological construction and specialisation of tools, as well as the occurrence of standardised (professional?) practices. These, in turn, can be used to infer craftsmanship and indeed trades focused on specific processing. Archaeological lines of evidence from the butchery record and indications of occupational topographies can be combined with historical accounts of particular building types, i.e. bathhouses, that indicate cultural attitudes to hygiene. Additional depth can be added to reveal ideologies as might be evident through depictions from religious iconography (Gould 1960). Religion can also be viewed indirectly. Lawrence and Davis (2011: 242) point to the presence of tinned fish and the relative absence of alcohol containers as indicative of adherence to Islamic edict by Afghan cameleers in Australia, the first Muslims to settle there, from the 1860s. Fish was not subject to *halal*, and thus tinned fish was a valuable source of protein. From this multidisciplinary perspective, we can go much further than an investigation of animal exploitation; the faunal evidence helps us to better understand community organisation (Ervynck 2004; Pluskowski 2010). Zooarchaeologists are increasingly cognizant of the diversity of social relationships that take place between humans and animals, and new approaches help to delve more deeply into these interactions.

Despite complications, the ubiquity of social division renders this topic of universal relevance for archaeology. Equally relevant are the ways in which zooarchaeology can provide insight into 'how' social divisions are manifested. There is far more to investigate than the ostracism of people associated with specific trades. Through constructed identities we gain insight into aspects of culture contact. Conquest appears to be a crucial link in the creation of outcastes. Outcastes often derive from indigenous communities, which are considered technologically, militarily, or culturally backward. The untouchables of India are thought to derive from remnant Dravidians conquered by northern Aryan groups and progressively marginalised into menial tasks (Gould 1960). This situation is replicated in Ethiopia (Todd 1978). Both these contexts also illustrate the physiological underpinnings of these social divisions; marginalised groups are invariably considered 'primitive'. In Japan, Eta apparently derive either from ancestral Negrito people from the Philippines, a Hindu tribe called Weda, or Korean immigrants (Price 2003: 41). The fact that endogamy is a common feature also speaks of the need to preserve cultural, physical, and moral purity. This theory of 'origins' has been much debated (De Vos & Wagatsuma 1966: 11–12; Todd 1978; Pankhurst 1999) but remains to be tested through archaeological data.

CONCLUDING COMMENTS

Whether discussing medieval Europe, Hindu India, or Shintō Japan, the act of killing animals, preparing their carcasses, and consuming their flesh has been associated with social or spiritual pollution. In contrast, medieval Europe offers both a counter and complement to these examples; there certainly were negative connotations, but these were not grounded in the act of slaughter per se.

In the European case, as part of the push for Western modernity, legislation to enact controls over physical pollution saw the gradual displacement of animal slaughter from the high street, giving rise to the abattoir. For the archaeological contexts, the attendant social facets that surround the crafts and materials we study are often elusive. However, this does not preclude us from using the rich ethnographic literature and modern analogy (Chapter 5) to inform our interpretation. Furthermore, the materiality of crafts can serve as a means of assessing technological developments (Chapter 6). Uniting the ethnographic and technological contexts is important in terms of recognising and capturing the entire scope of 'social influences'. For butchery, these are exerted on carcass processing itself but also exist as attitudes at a national level, which can result in both positive and negative associations for the occupational groups involved in the activity.

Scholars from numerous disciplines have demonstrated the ways in which butchery and butchering have had a profound impact on economy at a global scale, through meat consumption. Similarly, the universal and complex nature of social impacts has to be recognised. Acknowledging this point serves to temper and enrich our interpretations and connects us to the people behind the practice. It is interesting to consider that, as a consequence of ostracism and endogamy, there is an assurance to some degree that knowledge and techniques associated with the scheduled professions are transmitted from one generation to the next. The caste system effectively functions to embed geographic boundaries and practical skills in portions of society, particularly for those of a lower station. This phenomenon has been brought to our attention through anthropological enquiry. However, as the Canary Island example and cases from Tarbat and Toruń illustrate, they are also likely to be present in the archaeological record and could help explain important details of how craft knowledge is disseminated through time and across space, or why it is not.

Thus, whether from the perspective of localised ethnographies, systems of social differentiation based on associations with animal bodies, or newly industrialised settings where the butcher is effectively a factory worker on the lowest rung of the social ladder, there are opportunities to enrich our interpretations of the past. Once again, restating a point that is axiomatic for this book, the overarching emphasis on the study of meat and cut marks has circumvented our attention away from the craft and the practitioner, and this is to the detriment of our collective scholarly enterprise. Looking toward the future, this chapter has provided a starting point for imagining how a range of datasets could be judiciously weaved together to create a more intricate tapestry of past social attitudes to those involved in animal body-part crafts. This not only provides better ways to understand butchery and butchering in the past, it also uses cut mark data in more creative ways to better understand sociality. The role of the butcher is conceptualised in each society; whether or not it is acknowledged and articulated, no society remains ambivalent or ignorant of butchery.

CHAPTER FIVE

INTELLECTUALISING PRACTICE

Bridging Analogy and Technology

If ... by observing the adaptive behaviour of any living society, we can derive predictions about that society's discards, we are doing *living archaeology*.

(Gould 1980: 112)

'LIVING ARCHAEOLOGY'

The preceding chapter makes the case for a more comprehensive appreciation of the many ways in which social norms influence production, and how craft is enacted. In turn, this chapter draws the reader's attention to the type of knowledge – and practice – that constitutes butchery. The issue is a critical one for archaeology: how do we record, assess, and interpret experiential know-how from a position of inexperience? Put another way, how do we detach activity from phenomenology?

Addressing these questions is far from straightforward. Developments in *chaîne opératoire* guide us, at the disciplinary level, to recognize the diversity and complexity of layman knowledge and to make a concerted effort to bridge the gap between practical and academic know-how. Chapters 2 and 3 place the burden of evaluation on the cognitive underpinnings of processing and consumption – how and why we do what we do – rather than the materials and provenance of the artefact. This chapter emphasises the agency of the butcher, focusing on technology as a structural reference for the wider activities associated with butchering and for the study of cut marks. Here I adopt an autoethnographic approach, offering an account of my own experiences to reveal details that might otherwise be hidden from a zooarchaeological view of butchery. The aim is to better situate the specific type of practice we witness in the later historic archaeological record, focusing on commercial, systematic, and trade-orientated activity that is driven by standardised cutting techniques and specific tools. Although the examples I present are anecdotal, they illustrate contemporary correlates for transitions that occurred in the archaeological context. This includes changes in tools, demand, and attitudes to the craft, i.e., when butchery become professionalised. In this way, the discussion serves to rationalise and make sense of the global context, considered at a regional or national level, in order to position intensified scales of consumption, and indeed 'retail', as these may reveal themselves within archaeological contexts.

Through the principles of analogy, ethnoarchaeology, and experimental replication (Coles 1979; Mathieu 2002; Outram 2002), the goal of this chapter is to illustrate the constellation of activity, virtuosity of practice, and the diversity of knowledge bases that are inherent in butchery. Simultaneously, the following demonstrates that the butcher is familiar with and utilizes a wider gamut of knowledge systems than we might at first conceive. Modern analogy has played an important role in helping to fill gaps in archaeological knowledge of butchery (Thawley 1982; Peck 1986; Rixson 1988, 2000; Seetah 2008). From this basic premise, the chapter looks for continuities between modern and ancient practice. I also introduce and discuss the routes of specialisation in butchery. This is then consolidated in the next chapter through a detailed discussion of the materiality of butchery. Cutting implements in particular are an important instrument of butchery. However, situating the complexity of tool *use* into the overall schema of butchering requires a more extensive account of the activity that surrounds and influences the actions of the butcher.

At this point it is expedient to explain the time periods referenced in this chapter. 'Modern' refers to the twentieth century, in terms of the literature and research I discuss. I will also be including evidence from the Romano-British and medieval periods, where it is available, in order to draw connections with the material that is most relevant to Part II of this book. Certain of the following examples – for instance the discussion of waste – derive from both the modern context and historical research from the post-medieval period. The ensuing discussion does not generally engage with contemporary Western practices from the twenty-first century. Present-day industrialised and mechanised butchery in the Western world bears little relation to the practices we observe in the archaeological past, and I have tended to avoid using analogy from the contemporary industry except in specific cases.

Furthermore, in the following discussion 'Western' is not used in a geographic sense, but encompasses Australia, South Africa, and South America, not only Europe and North America. In fact, Eastern Europe and the Mediterranean do not fall within this categorisation, considering the different types of cuisine from these parts of Europe. I make this distinction based on a view of meat exemplified by a 'steak': a single portion that forms the central feature of a dish, usually fried or grilled. This form of retail and preparation is generally restricted to specific regions of the world where both husbandry and butchery are tailored for the production of this 'cut' (Swatland 2000). Meat as steak depends on the animals' constitution; the evidence indicates that from the Iron Age to at least the Middle Ages in Britain, meat was generally boiled, suggesting the animal's flesh was tougher (Rixson 2000: 122–3). The topic was treated in an insightful and thought-provoking film *Steak (R)evolution* (2014), viewed from the perspective of chef Yves-Marie Le Bourdonnec.

DRIVERS OF MODERN BUTCHERY

Building on the position developed in Chapter 3, outlining the milieu within which a 'butchery ecosystem' is situated, here I focus attention on the 'working mind'. A butcher will utilise a detailed operational knowledge of the physiology of different animals to deal with skin, muscle, tendon, viscera, blood, and bone. Paradoxically, an animal's morphology is both a mechanism of constraint and a facilitator to the butchery process. *Pressure* is applied at the joints to disarticulate bones from their sockets; the ends of limbs are used to provide *leverage* to expose parts of the carcass to the knife that might otherwise be hard to access; *gravity* is relied on to facilitate evisceration and exsanguination; precise physical *manipulation* of the carcass is called for to position and reposition the animal during butchery. Principles of biology are well understood and consistently used, for example, dealing with and mitigating the processes of spoilage and suggillation.[1] The butcher exercises control over these factors. However, we also need to consider those parameters that fall well beyond the purview of the butcher, for example, the environment. Different ecological settings will directly influence the way animal body parts are treated (Mowat 1970: 129); as the vignette discussed later from Kimana will illustrate, ecology also plays an important role in dealing with waste. In the absence of refrigeration, environmental conditions and climate would have been particularly important in the past. A recently killed carcass undergoes *rigor mortis*. Rigor mortis plays a major role in the timing and process of butchery and is directly influenced by ambient temperature (Gerrard 1964: 237–8). This illustrates the relationship between time, biology, ecology, and butchering.

By concentrating on a suite of factors that have shaped modern butchery, the following provides a deeper understanding of the mechanics of how the meat trade functions, as well as details of professional practice. I deliberately focus on facets of the meat trade that are less well recognised but arguably more relevant to zooarchaeological studies of butchery, as they have played a defined role in shaping cutting techniques. In particular, two topics, the control of waste and the evolving concept of aesthetics in meat presentation, emerge as particularly relevant to archaeo-historic enquiry.

Waste and Its Implications for the Meat Trade

The following takes a broad look at animal waste within the urban environment, aiming for a more informed assessment of how inhabitants of large settlements and societies dealt with, and were influenced by, the unwanted derivatives of carcass processing. In this way, the text brings aspects of butchery to our attention that we could potentially observe in the archaeological record and which influenced the creation of the assemblages we study. Acquiring a more complete picture of waste – what it is, practically and metaphorically – enriches our interpretation of archaeological butchery.

The concept of waste in archaeology is complex (Dobney et al. 1998; Bartosiewicz 2003). The following examples refer specifically to one type of waste, the discard from carcass processing. However, 'butchery waste' is itself a complex issue and covers waste products, i.e., effluent, as well as wastefulness (in the sense of injudicious use of resources).

The latter point is important, as perceived wastefulness might be misidentified as inefficient or as evidence of crude practice. Indeed, the modern industry provides numerous examples of activity that might be considered wasteful to a non-specialist but actually has more to do with market strategy than with poor technique or inefficiency. For example, modern meat retail might be considered wasteful due to the variety and volume of animal parts that are discarded or not offered for sale. However, many carcass portions are rejected due to potential risks to human health; in the EU the cerebral and spinal material of cattle are removed and discarded at the abattoir, following the bovine spongiform encephalopathy scare that surfaced in the late 1980s in Britain (Greenlee & Greenlee 2015).

Saleability is another primer for selective retail, which in itself can serve as an index of a society's overall view of carcass-part utility. More cynically, certain parts are not offered for sale in order to maintain the value and prestige of other portions. These portions, often described as 'noble cuts', can have a morphological basis for differentiation and may derive from parts of the animal that are less likely to be exercised during rearing, such as the rump or loin (see the noble cuts described in Ranken 2000: 3–6). In contrast, cuts of meat from the neck or shoulder, which are more likely to be exercised and therefore will have stronger connective tissues, are considered less noble. While we might assume that carcass divisions such as these, based on ideology, are a feature specific to the modern industry, this is not the case. Ethnographic studies across many contexts show that particular parts of a carcass will be discarded in order to satisfy ritualistic norms, as well as notions of sanitation and hygiene. The Huaulu, of the Moluccas Islands, Indonesia, for example, follow this protocol, discarding viscera prior to bringing a carcass into the settlement (Valeri 2000: 314).

As a starting point, one of the more pertinent questions is: what actually constitutes 'animal' waste? If we approach this from an archaeological perspective, the answer is simply 'bone'. In this case, bone is waste within the context of discard and a by-product of consumption but can certainly have many subsequent uses, and indeed biographies, as a raw material for the production of tools and ornamentation.

However, for the industrialised West, bone has increasingly become normalised as a waste product in a nutritional sense. This scenario is only possible where market forces exist that place a greater monetary value, and aesthetic emphasis, on meat that has been deboned (Seetah 2008). Deboned beef has become the norm, with bone removed from the meat soon after slaughter, producing 'primal cuts' (Oliphant 1997). Butchering has influenced this transition. The reorganisation of the activity into a highly specialised craft, removed from public view, serves to distance the general population from any contact with the processing of the meat they consume. Thus, numerous components of an animal are now considered as discard. Alongside bone, most internal organs and the animal's blood are largely viewed as waste products, at least within the context of commercial sale to the public.[2]

This was not the case historically or archaeologically, nor is it the case in many contemporary societies, simply because there is much nutrition to be gained from bone in the form of marrow and grease. Bone would have been sold with meat and would only be discarded after cooking and consumption. Therefore, it is perhaps more appropriate to consider waste within the context of carcass by-products. This allows us to separate bodily waste (urine, faeces) from non-meat tissues that are considered waste from an economic viewpoint.

When considering waste produced by the processing of carcasses specifically, it is as settlements transitioned from low-intensity (rural) to high-intensity (urban) settings that dealing with effluent became increasingly problematic. Until the early modern period, the slaughter of animals in urban settings generally took two forms: commercially, on the butcher's premises and usually in a separate room, or as part of smallholder husbandry, taking place in back alleys. In both cases, the effluent would usually be washed into the street, as documented even in late eighteenth-century Paris (Watts 2008: 13). Slaughtering and disembowelling a cow results in a large residue of urine, faeces, and blood. Performing these tasks on the butchers' premises led to a number of issues related to health and nuisance. As early as 1361 in London, this 'caused an abominable stench which poisoned the air of the City, while blood flowed down the streets' (Jones 1976: 78). Access to water for washing the carcasses also became critical. In the absence of piped mains water, a solution was to group the flesh trades together, located close to a source of water (Goldberg 1992: 64–6; Yeomans 2007). An alternative was to cart waste to watercourses to be jettisoned (Jones 1976: 78–9), a principle that also applied to tanners and other forms of artisanal and domestic waste (Metcalfe 2012: 99n2; see also Day 2005: 181 for a historic example from New York).

The management of waste from disembowelment necessitated special attention, particularly since in Europe this was thought to contribute to miasma (Perren 2008: 124). Control of waste, irrespective of location or settlement size, was closely linked to intensification in carcass processing activity. The issue of scale has a number of ramifications. The meat trade, alongside fishmongers and poulterers, contributed to animal waste in the urban setting. However, the volume of waste derived from mammalian carcasses was significantly greater than that from either poultry or fish. The increased demand for meat in urban enclaves, which occurred as a consequence of an increasingly centralised population (Forshaw & Bergström 1980: 47; Watts 2008), was also a major catalyst for changes in waste management. From 1550, butchers in London, actively seeking to increase their share of 'flesh sales', created waste on a scale that galvanised the transition to controlled disposal (Jones 1976: 122–4). This also led to new legislation and had numerous implications for urban planning.

In Europe, a continuum existed between rural and urban disposal up to the post-medieval period, the point at which the abattoir, a designated slaughter-house and a purely urban entity, altered the architectural landscape of many large metropolitan centres (Loverdo 1906: 58; Lee 2008: 46–7). The development of the municipal slaughterhouse has received relatively little attention, despite its significance to modern-day life and, indeed, modernity itself. Young Lee notes that 'historians have largely overlooked its [the abattoir's] impact as an urban phenomenon susceptible to cultural analysis. Created to meet the increasingly large and geographically concentrated urban populations, the slaughterhouse is a social instrument responding to the demands of a gargantuan belly' (Lee 2008: 2). The abattoir was seen as an advancement of industrial and architectural design, encouraging the commodification of animals and their parts (Corbin 1995: 178).

While this was hardly a novel concept (Seetah 2005: 6), it had never before evolved into such a refined and ordered system. It can be argued that the development of the abattoir, alongside rapid transformations in the butchery process itself, led to a transition in the whole ethos of meat trading. Animal slaughter and the disposal of effluent were summarily compartmentalised and detached from the rest of meat processing and retail (Lee 2008: 4; Fitzgerald 2010). This in itself was a consequence of a more fundamental problem faced by meat purveyors: the requirement for fresh meat (Otter 2008: 89).

Legislators dealing with waste management could have required butchers to bring in fresh produce from outside the city, but this had two problems. First, a side of beef would require up to four horses to transport it to a city, whereas the live animal could transport itself (Forshaw & Bergström 1980: 60). Second, longer transportation distances increased the likelihood of the meat spoiling. Thus, the processing of animals had to be pushed to the periphery of the city, but near enough to minimise economic loss. Centralised processing and distribution became the norm (Lee 2008: 3–4). Live animals were no longer brought to a meat market and sold for subsequent processing. Instead, meat was distributed to the point of sale, which meant that the problems of waste management were confined to specific locations.

As this system developed, rail transportation became the mainstay for bringing animals to the abattoirs. By the early twentieth century in Europe, abattoirs had transitioned from being relatively small scale and deliberately nondescript to huge constructions that had their own freight stations and holding pens that could accommodate thousands of cows, calves, and other domesticates (Lee 2008: 51–65). Figure 5.1 illustrates the '*Marché aux bestiaux et Abattoirs de la Mouche*', which was the vision of architect Tony Garnier and came to be a central feature of the city of Lyon, France (see also Lee 2008: 64–5, for a detailed description of the development and functioning of this abattoir).

Economic growth centred on animal products, increasing requirements for quality and waste control, the growing influence of the guilds, and a humanitarian outcry on the treatment of animals led to the establishment of the abattoir system. In combination, these factors were instrumental in turning the meat trade into an industry. Viewed from this perspective, one can appreciate that the need to remove waste, odour, and death from the city indirectly led to fundamental changes in meat processing. These factors galvanised the institutionalisation of slaughter, which now exemplifies our own mode of meat retail. An outcome that would have been hard to predict is the increased detachment from production and greater levels of compartmentalisation of animal bodies and butchery. 'What had disappeared, though, was the quotidian encounter with slaughter rather than the act of slaughter itself' (Lee 2008: 51). This model paved the way for intensification in meat retail, which has come to be heavily reliant on aesthetics and representation.

However, before discussing aesthetics, and as a way to add nuance to the archaeo-historic case studies, I augment the description of 'task groups' in Chapter 3 to describe the management of effluent as noted in a non-Western ethnographic context. The case is a revealing one, as it provides an insight into different scales of waste control. Four tiers of carcass processing are practiced in modern-day Kenya: around the homestead, for example as performed by the Maasai within their village (Chapter 6, see section on Archaeological

FIG. 5.1 Aerial image of La Mouche, a slaughter-house in Lyon, France. (From T. Garnier (1921), Grands Travaux de la Ville de Lyon, with permission from the Manuscripts Room, Green Library, Stanford.)

Relevance); in a 'slab', which is a basic slaughterhouse that can accommodate the slaughter of about 40–80 animals per day, almost exclusively consisting of cattle, sheep, and goat; and at a slaughterhouse proper, a larger-scale, better-equipped establishment with the capacity to process larger numbers of animals, in the hundreds per day. For both the slab and slaughterhouse, slaughter and butchery usually only occur on a set day in the week, depending on demand. Thus, Wednesday is the usual day of operation in Kimana and the surrounding area, as the cattle market operates on Tuesday. The final and most elaborate and advanced establishment is the abattoir, one of which is situated on the outskirts of Nairobi and supplies meat for both the capital and for export.

Taking the operations of a slab as an example (of which I visited three, one each in Kimana, Emali, and Mbirikiri), the control of waste follows a set pattern. Following slaughter, the concrete floors of these small-scale processing plants – the 'slab', hence the name – facilitate sluicing of blood into a system of channels (Fig. 5.2a). The concrete gutters direct the blood to a rudimentary collection point, where it is collected in buckets and poured into a containment pit to decompose (Fig. 5.2b). Faeces extracted from the intestines during cleaning are first collected in wheelbarrows (Fig. 5.2a), and then deposited in an open area, which progressively builds into a large mound as more waste accumulates. This effluent is allowed to dry. This system of waste disposal raises a number of questions that resonate with archaeological enquiry. For example, when does one system (localised disposal) give way to another (centralised)?

FIG. 5.2a Kimana slab, Kenya: gutters for channeling blood to a collection area. (March 2017, by author.)

FIG. 5.2b Kimana slab, Kenya: containment pit for accumulated blood. (March 2017, by author.)

FIG. 5.3 Dung accumulated over several years, from both slaughter and when animals are contained in enclosures. This is then collected by local agriculturalist and used as a fertiliser.

What is the significance of the ecological context? Despite close proximity to dwellings and other businesses, due to the intense heat, blood and faecal waste quickly dries. There is little discernable odour in the surrounding environment; these practices do not initiate a sense of disgust in the local population, nor is there a sense that the slab is creating a nuisance.

What are the uses of effluent in circumstances where resources are more extensively exploited? The disposal of faecal waste in the slab is identical to how disposal takes place among the Maasai, at the smaller-scale example of a homestead. In the Maasai case, faecal waste is collected in the kraal (an enclosure for cattle) and allowed to dry (Fig. 5.3). Once desiccated, it is scrapped and removed by local agricultural tribes, such as the Kikuyu, who use the dried and nutrient-rich dung as an arable fertiliser. This last point illustrates that many by-products of slaughter become waste only when production outstrips demand.

Aesthetics

Representing flesh and fauna in specific ways, driven by both economics and cultural ideals, has a long precedence.[3] During the medieval period in Europe, the display of meat and game was a crucial aspect of banquets and feasting (Lehmann 2003). The visual stimulus formed a central part of the event, serving to augment the ceremony and pageantry that accompanied consumption. While the connection to butchers and butchery is not straight-forward (large estates, for example, would use itinerant butchers far removed from commercial practice; see Woolgar 1999: 114), the appearance of the meat, and need for a distinct mode of presentation, had implications for the butchery process. Furthermore, in addition to representation in a feasting context, aesthetics as an expression of quality played an important role in meat trading during the medieval period (Rixson 2000: 120).

The modern iteration of aesthetics can be regarded both as a standardising visual element, a driver for turning a carcass from flesh into 'joints' and boneless portions – into meat – and as an agent promoting saleability (Mancini 2009: 89). Presentation of meat is a complex aspect of butchery (Mancini & Hunt 2005). Our exposure to meat aesthetics is ubiquitous. It is almost impossible in Western day-to-day settings to purchase meat that has not been standardised in terms of portion size and packaging. In the modern setting, the representation of flesh is manipulated to leave a strong visual impression. In addition to driving standardisation of portion size, aesthetics serve as a visual stimulus to promote ideas of meat quality, based on industry standards (Gerrard 1964: 113–30; Elmasry et al. 2012).

The modern perception of 'good meat' – a subjective term – centres on the vivid red and pink colouration of the flesh. This is combined with a relatively minimal amount of visible fat that is localized around the muscle, not inter-spersed in the tissue itself, and is therefore easily removed; compare, for example, Western ideals versus those from Japan, which are typified by a very high degree of marbling. Finally, 'good meat' rarely contains bone. Accord-ingly, when meat is presented for sale, as meat cuts, it is usually under lights that enhance the red and pink tones of the flesh, with the majority of the visible fat removed, along with any bone. It is probable that societies in the past, as with many modern cultures, viewed a carcass as providing meat and not specific cuts.

Staying with the modern Western case, bone features strongly in terms of the physical activity that drives the aesthetics of meat retail. Bone may be required to return a profit as it provides added weight. An animal that carried relatively little meat compared with bone would not prove profitable if deboned, as meat is generally the higher-profit commodity. Bone can be sold separately but does not usually generate the same income. Therefore, rabbits are usually sold either whole, skinned and beheaded, or portioned, but essentially as an entire unit with bone.[4] In this scenario, one must also factor in the time spent on such an activity, which has implications for the overall efficiency of the business. Thus, carcass aesthetics depends on the anatomy of the animal, a point I will revisit below. An animal has to be suitably 'improved', carrying adequate physical flesh, to make deboning practical and profitable.

In the same way that the need to manage waste effectively led to the abattoir system with attendant techniques adapted to meet a new approach to

FIG. 5.4 Cutting chops between individual ribs. Before the advent of the band saw, such techniques were primarily driven by the morphology of the animal, as cutting between the ribs avoided bone. More recently, the desires to minimise visible bone and to standardise portion size have become increasingly important drivers of butchery.

provisioning, so too has the requirement for aesthetics led to new cutting practices. In the modern Western setting these include increased deboning, particularly of larger animals like cattle (Gerrard 1964: 123). There is also a marked increase in the use of the saw (and more recently, the electric band saw). In part, this is to deal with the larger animals of the modern day, but it also helps mitigate the splintering of the bone and increase uniformity of the final product. This has both culinary and aesthetic drivers. For the former case, broken slivers of bone are universally avoided in modern Western cuisine. Use of the saw plays a role in presentation. The bone itself, when cut with a saw, remains flush with the meat. If chopped, it splinters and is more prominent within the surrounding flesh. Use of the saw therefore minimises the distinction between bone and meat. If the cut includes bone, the visibility of the bone is minimised through specific cutting techniques. For example, cuts will be made at an obtuse angle, rather than perpendicular to bone. This has the effect of tapering the portion of meat, extending the fleshy part, so that it appears relatively larger than the bone component. This satisfies economic and aesthetic drivers: the bone is integrated into the meat, but with an impression that there is relatively more flesh than bone. Chops, which include part of the rib and vertebrae, are cut between rib bones (Fig. 5.4). This is primarily driven by the skeletal morphology, but it also serves to minimise the visible bone on display. Thus, presentation depends closely on the relative proportions of meat and bone, not only from an aesthetic perspective but also from an economic one.

While aesthetics must seem an innocuous aspect of meat retail, it actually illustrates the wider influence of visual representation. From the tissue of the animal to the processing of its flesh and finally the act of cooking, all forms of presentation are critical. The aesthetics of animal representation, for example during medieval feasting, are well documented (Lehmann 2003). However, the broader topic would benefit from a deeper interrogation, particularly considering its importance within the schema of meat consumption, retail, exchange, and sharing. Above all, the modern setting signals that meat and bone need to be considered within a more cohesive framework, either drawn from a better assessment of cutting practice from archaeological contexts or from complementary sources of evidence.

SPECIALISATION OF TECHNIQUES USED IN BUTCHERY

Despite the fact that an essential thread connects the earliest butchering to the modern iteration of this activity, there is clear evidence for transitions in how butchery has been undertaken. As such, the techniques of butchery are themselves a dynamic expression of agency. Through time, we observe that the animal's skeletal morphology becomes less of a constraint. Changes in techniques of butchery are driven by transitions in implements, particularly with developments in tool functionality and durability. While the implications of the evolution from lithic to metal implements on carcass butchery is easy to appreciate, the development of large versus small tools, as well as heavy versus light and thin blades, have all played a role in how the techniques evolved; these are the technological underpinnings of butchering. These features can be thought of as internal to butchery.

Economic drivers that influence techniques per se are evident at a broader level. Butchery developed from an innovation to a skill to a craft; with increasing specialisation and uniformity, butchery transformed into a trade and ultimately an industry. This schematic, while helpful, grossly oversimplifies a complex trajectory. The connection between the developments in butchery techniques and society at large can be generalised diachronically. The evidence suggests a progression from small-scale, rural, and idiosyncratic, e.g., the British Iron Age, to large-scale, urban, and standardised practices as witnessed on military and urban sites from the Roman period, culminating in a modernized, industrialised activity, reliant on mechanisation, in the contemporary setting (the complexities of this trajectory are more nuanced; see Chapter 11).

The above emphasises a specific view of technology, one focused on materials and sequences of production; it borrows heavily from *chaîne opératoire*. Before discussing specific techniques and principles, and in the interests of situating butchery in a more accurate, global position, a number of additional associated technologies need to be introduced. These include the development of methods, techniques, and paraphernalia for mid- and long-term storage, which formed a natural progression from more immediate consumption.

Larger and varied sets of technologies connect butchering with animal husbandry. Of particular relevance is the 'improvement' of an animal's constitution through selective breeding, as well as agricultural developments (Thomas 2005; Davis 2008). In one sense, biology is being driven by the desired outcomes of the retailer and consumer and enacted through butchery. In effect, the latter point served to place more animals into networks of consumption, while the former resulted in larger quantities of meat. In both cases we effectively have more to process: larger individuals in greater numbers. This had a direct impact on techniques of butchery.

Finally, serving as a nexus between technology and social factors are the ways in which different modes of cuisine have influenced butchering. This point is rather more complicated if viewed over a long span of time. Butchery must have initially been driven almost entirely by cuisine and other sociocultural factors, in addition to the basic desire to exploit a source of nutrition. However, by the Iron Age, when we can consider a craft to have come into existence (Grant 1987), and certainly by the time we see the birth of a trade in meat in the Romano-British period (Maltby 1989), we effectively see two distinct modes of butchery emerge, both revolving around cuisine. In one scenario, carcasses are butchered and animal nutrition is compartmentalised for sale. In itself, this may involve one set of practices and practitioners who slaughter, flay, and portion the carcass, followed by further reduction by another set of practitioners. The meat is then sold. Proceeding from retail, another mode of butchery takes place at the household level: kitchen butchery ('reduction to household or pot-size pieces': Rixson 1988). At each stage, different techniques are in employed.

Chaîne Opératoire *of 'Deconstruction': Techniques and Processes of Butchery*

As I began to elaborate in Chapter 3, butchery represents an idiosyncratic example of 'production', one in which the creation of a product for consumption – meat – mandates the deconstruction of another product from farming or hunting – an animal's body. Numerous authors have emphasised the sequential of nature of butchery, outlining the stages of slaughter, evisceration, and skinning through to final cutting and jointing (see Lyman 1987: 253, table 5.2, for an overview of studies on processing). Generalisation and simplification into discrete classifications is necessary and useful. However, one cannot lose sight of the fact that butchery, like other forms of productions, is a process that has almost limitless inherent variability. While it is impossible to discuss every single permutation for each cut mark noted from the archaeological record – to identify in detail the entirety of each mark's biography – the following provides a number of guiding principles taken from modern contexts that could help to inform archaeological interpretation.

Well before any butchery has occurred, the choice of tools and location for processing are often culturally and economically meaningful. As noted from historic and contemporary industrial literature (Jones 1976; Billington 1990; FAO 1991), speculated from the archaeological record (Birley 1977: 40; Maltby 1979), and observed ethnographically (Mooketsi 2001), slaughter invariably takes place in a specified setting. In Botswana, for example, the kraal is considered the proper place for slaughter, as it is the home of cattle (Mooketsi 2001). Each group in Mooketsi's study favoured a different implement for slaughtering. The Bakwena employ a gun, the Barolong utilise a sharp knife, and the Hambukusha an axe. Pragmatism plays a role in choice of location. The Bakwena indicated that slaughtering in the kraal avoided the risk of stray bullets hitting bystanders. If possible, the distance between place of slaughter and subsequent redistribution is also minimised. Humane treatment of animals is considered crucial if the quality of the meat is to be maintained following slaughter (Rixson 1988). Animals are not usually fed in the twenty-four hours proceeding slaughter but are given water. Stressing the animal is avoided[5] (Ashbrook 1955). Knives are sharpened, and if required, hanging paraphernalia are assembled (Mettler 1987: 1–7), demonstrating the degree of complexity involved in preparation for butchery.

The act of slaughter itself takes varied forms. Slaughtering techniques are instrumental to the conceptualisation of butchery as a culturally salient performance. In religious situations such as *halal* butchery, both the carotid and jugular are severed in order to maximise exsanguination (Rixson 1988). Kosher practice relies on a single slanted cut to the carotid, rather than slitting the throat. This served cultural ideals, as it distinguished Jewish practice from the method employed by the 'gentiles' (Cope 2004: 26). The kosher technique also had practical benefits. It promotes bleeding and minimises 'the bread of strangulation'.[6] Jews avoid this by-product of slaughter, which results from the animal choking on its own blood when the throat is slit (Cope 2004: 26–7).

Slaughter is a singular act, the point at which life is extinguished, along with consciousness and sentience. Modern analogy and ethnographic examples illustrate the lived experience around the act of slaughter, which adds richness for an archaeological audience. The modern context also draws our attention to the fact that the complexity and nuance that surrounds slaughter extend to all facets of butchery. However, the archaeological record mainly offers bone, as a source of evidence for this aspect of human interactions with animals. How can we improve interpretation from the archaeological record by engaging with the lived context?

Approached from the perspective of *chaîne opératoire*, cut marks reveal the details of causal agency. They provide evidence of the techniques involved in the dismemberment process. *Techniques* drive the placement and manner in which cut marks are made. For example, marks noted on the posterior surface of cattle metapodials have been used to suggest that the animal was suspended, at least during the early stages of dismemberment (Hambleton & Maltby 2009). Alternatively, marks on the anterior surface have been interpreted as evidence for butchery taking place while the animal was lying on the ground (Wilson 1978: 122; Maltby 1985a: 23). In essence, the difference in technique relates to the part of the animal exposed to the butcher. Hanging the carcass brings the back of the limbs to hand; this scenario is reversed when the animal is lying on the ground (however, variations do exist, and the above should not be read as static rules).

Techniques depend on a number of factors, in particular, the implement and how it has been employed. Is there a propensity toward knife cuts as opposed to cleaver chops, as noted from the British Iron Age, versus the Romano-British period (Grant 1987)? Are the cuts themselves prominent? This may indicate the presence of specialised cutting tools, the likes of which are recovered from both Roman and medieval British sites (Manning 1979; Cowgill et al. 2001; see Chapter 6). Adopting an approach that is well established in other material specialisms, the depth and size of the cuts indicate a specific 'overall technical style' of butchery (as noted for pottery; Smith 2000) and within broader regional debates (Gosselain 2000). A collection of short, fine marks, with consistent geometric clustering that indicate skinning, jointing, and meat removal activity, would imply a technical style dependent on smaller knives for the entire disarticulation process. From this it would be possible to estimate the speed with which the whole process was performed. Alternatively, small knife marks at skinning sites, cleaver marks (chopping cuts) through the head of the femur bone for jointing, and large blade paring for meat removal – which characterise Romano-British practices – would indicate a completely different technical style.[7] This latter scenario points to the transition through a range of implements and may even suggest that the trajectory of processing was compartmentalised into stages that involved different craftspeople.

If cut marks are the outward manifestation – the outcome – of the butcher's actions, techniques are the means by which the butcher achieved the outcome; they are the skilled expression of 'how' the butchery was performed, of the individual steps within the wider schema of the entire underlying mechanism or *chaîne opératoire*. Cut marks are thus dependent on the techniques. But how, in this process, can we identify craft? This is the point at which individual processes become important. For example, skinning, jointing, and meat removal are all individual stages. However, the processes involved go beyond this. Each constellation of activity, from how the animal is killed to final disposal or movement of the bones to other locations for working, forms its own process. The specific cuts that constitute a given technique for removing the skin from the animal, for example, come together to create the *process* of skinning.

Analysing butchery in these increasingly complex ways helps to reveal behaviour. Processes are not static, and do not necessarily follow a logical path. They expose variation that can be important for interpretation. For example, the outcome may be clear, i.e., there is evidence for disarticulation of the leg at the femur through cleaver chopping. However, it may not be possible to establish the series of actions and decisions that occurred to arrive at this stage, nor the order in which they occurred. We may speculate, logically, that the stages would follow a progression, from skinning, followed by evisceration, and then removal of the hind leg.

However, without evidence for skinning or evisceration, the analyst would have to acknowledge that these stages may not have occurred or are unrecognisable from the archaeological record. The butchery 'strategy' might have dictated that only the highest meat yield portions were required. If so, there would have been no need to skin and eviscerate the animal when it would have sufficed to simply disarticulate the meaty portions and discard the rest. The resulting bone assemblage would typify a 'kill site'. This profile would not characterise the strategy utilised at a processing site or an urban Romano-British settlement.

While it may not always be possible to establish the exact route taken to achieve an outcome, this is not necessarily detrimental. Such a scenario may lead to interesting interpretations regarding why certain stages in an operational chain seem to have been omitted and/or may provide a view of the options that were available to the butcher. This point reinforces the utility of assessing the entire process, where possible, in order to visualise the interdependence of individual tasks. A final point to note from this example is that it illustrates the limitations inherent in the archaeological butchery record, signalling ways in which analogy may help to fill gaps in our interpretation.

APPLYING MODERN ANALOGY TO ARCHAEOLOGICAL INVESTIGATION

This chapter has taken cues from the modern industry to illustrate how waste and quality control, as well as the technical aspect of butchery itself, resonate well beyond the context of meat production. The management of effluent in particular was and remains an issue with far-reaching sociopolitical and ecological ramifications. But can these examples help answer archaeological questions, chiefly, how can we better assess the wider context that surrounds production and consumption?

The transition to the abattoir system represents a specific and defined development in meat trading, which led to a rapid and comprehensive transformation in cutting practice. In terms of the implications for how meat is produced, the magnitude of this transformation is comparable to any major transitional development within an archaeological context, e.g., the advent of agriculture and development of new systems of animal husbandry; a major episode of conquest; or the replacement of one religious ideology with another. In all of these examples, the result was likely a defined change in how humans turned a carcass into meat, whether in practical or ideological terms, or both.

The changes that took place following the move to the abattoir system offer points of resonance, in principle at least, with developments that we may be able to observe in the archaeological record. For example, there is a comprehensive transformation in the process of butchery. This is characterised by a factory-line approach, where individuals undertake a precise and specified action, effectively resulting in a near-complete dissolution of 'technique'. The taskscape is fractured into a wide mosaic of discrete acts. Alongside suppressing the inherent craft, a marked transition in 'who butchers are' is also observed. A recognised profession as part of a trade gives way to an industry reliant on cheap, unskilled, often migrant (or other marginalised group) labour.

The ideology of meat aesthetics, which extends back at least to the Roman period in Europe and likely much earlier, forms a step along the process of standardisation. Initially, goods would have been bartered: a shoulder of mutton in exchange for a quantity of grain. Once systems of measurement were in place, the retail of meat and produce would have been by weight. This culminates in the modern day with standardisation of portion size serving as a reference for both amount and quality.

However, another facet of meat aesthetics connects the modern context much more concretely with the archaeological: the development of butchery techniques to process carcasses to improve their visual appeal. Modern Western butchery emphasises deboning prior to cuisine, not after. Perhaps most important from an industry perspective, as much of the meat as possible is sold as steaks or roasts. Steaks and roasts are cut from all parts of the animal, some of which is sold at a premium, most of which has little if any bone. For the modern case, the remaining portions of the carcass are considered inferior. The flesh is still meat, but it cannot be cut and sold as steak. It will be cubed for stews, minced, etc., and commands a lower price. Furthermore, cubing and mincing allows for meat from different individual animals to be sold as one unit. Buying a steak or roast invariably ensures that the portion of meat derives from one animal, and one part of that animal. This is important when considered from the perspective of modern cuisine. Preparing a roast requires consistency in constitution to ensure uniformity of the cooked product.

Approached from this perspective, one can appreciate that butchery, retail, and cuisine develop around a basic premise of animal constitution. From at least the Roman period in Britain, morphological improvements progressively modified the constitutions of traction animals (Albarella et al. 2008). By the post-medieval period, with intensified demand for meat, selective breeding of larger animals led to new breeds raised solely for meat (Ritvo 1987: 45; Mac-Gregor 2012: 430). In essence, animals were customised to meet new economic requirements. This illustrates two levels of aesthetics: the first that acts on the gross morphology of the animal, and which effectively anticipates the second level, that which will occur for meat retail, through butchery. Accordingly, aesthetics in this regard not only references 'presentation' but also the underlying, and increasingly exaggerated, commodification of the animals' bodies. Husbandry progressed from producing animals that provide greater traction

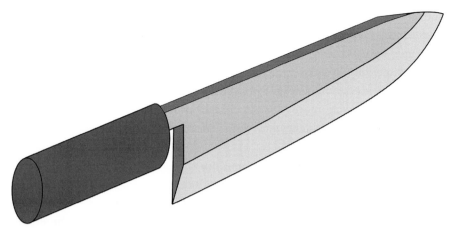

FIG. 5.5 Blade with an edge on one side only, typified by Japanese-style knives.

power to animals that had increasing specificity in their underlying morphology. Ultimately, this led to the modern aesthetic ideal: lean red meat in the form of a steak, with visible but moderate marbling and little if any bone.

The influence of tools is discussed in greater detail in Chapter 6, but a note worth making here is that whether discussing the medieval, post-medieval, early modern, or contemporary context, the relationship between tools and aesthetics is an important one. An example of this can be seen with 'single edged' knives (Fig. 5.5), which cut on one side of the blade, and with steel included in that region only (Nozaki 2009: 16–17). This minimises drag, producing a smooth cut, and is considered essential for fish destined for sashimi.

In summarising the main points from this chapter, three expectations can be put forth regarding the ways that the modern context can help us better assess the archaeological record.

The ensuing text should be considered within the context of a process of reflexive negotiation between 'bone' from archaeological contexts and 'meat' from a modern butchery perspective: which cut marks or characteristics thereof (from a contemporary context) indicate a shift in demand, specialisation of tools, or need to expedite the butchery process?

Expectation 1: Demand drives butchery.

In the modern context, this can be observed in the type, uniformity, and relative proportions of marks. The modern example also illustrates that, despite tools that facilitate butchery, the techniques themselves can become more physically demanding, placing greater strain on the butcher. In this regard, the changing constitution of the animal, becoming larger with denser bone, contributes to the increasing physical burdens of the profession. For the archaeological context, sites where levels of demand increase, e.g., urban and military Romano-British enclaves compared with rural, or large urban medieval centres compared with Romano-British villas, will show evidence that resonates with what we see today. Transitions in the archaeological record rest on observing improvements in implement technology, such as steeling of the cutting edge, coupled with the creation of specific butchery tools (see Chapter 6). Having more durable and heavier tools, which can generate more momentum, facilitates the cutting process, but would have been more physically demanding.

In addition to an increase in proportions of cut marks, intensified demand leads to a shift in the type of cuts. Changes in technology, observed in new tools and techniques, catalyse modifications in the way tools are used, where on the carcass they are used, and frequency of use. An example of this comes from the Romano-British period, where the cleaver supersedes the knife as a tool for disarticulation at the main long bone joints (see Chapter 9).

Expectation 2: Morphology drives butchery.

This expectation is easier to rationalise considering the obvious variations in constitution across a carcass. However, there are nuances to consider. Looking at the carcass in units, different techniques and cut mark frequencies are required to accommodate bones that form close and regular articulations, such as ribs and vertebrae, as opposed to the long bones. Variability in soft tissue and the specific musculature, in terms of size, ease of removal, and the way the muscle is attached, will drive specific aspects of butchery – e.g., do tendons need to be detached? Is the muscle-to-bone attachment particularly robust?

Taking individual bones in isolation, variations in bone architecture will drive differential cutting activity for meat removal; for example, compare the fore with the hind limb. In the former case, the scapula, which is the main bone that has to be negotiated, is positioned within a large muscular mass of flesh, but is awkwardly shaped. Thus, more cuts and time are needed to debone the shoulder area. In contrast, for the hind limb the large muscles that form the haunch encase the femur, which is relatively straight in shape and easily removed from the surrounding flesh. Furthermore, the proportion of cancellous (soft, spongy) to cortical (hard) bone plays a role in how the bone is processed. For example, where cortical bone is thickest, such as long bone shafts, marrow extraction relies on splitting the bone – with chopping cuts – either transversely or longitudinally. Elements that are predominately composed of cancellous bone will still require chopping but can be processed in a less specific manner, although this may still be systematic (Fig. 5.6).

The anatomy of the animal drives butchery in numerous other ways. In the modern context, the processing of larger animals (but also the influence of

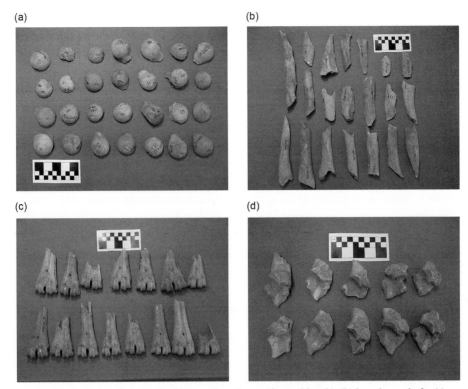

FIG. 5.6 Fracturing techniques for (a) cancellous bones (femoral heads); (b) long bone shafts; (c) dense cortical bone (metapodials); and (d) complex joints (calcaneus bones from the hock joint).

aesthetics) has caused a shift from using knives to saws and the contemporary electric band saw. Hanging paraphernalia has also become an essential part of modern butchery, as indeed has mechanisation. The result of these collective changes is that, while many more individuals are involved in the industry due to increased demand, any one person can effectively perform with greater efficiency. One worker can hook lifting apparatus to the Achilles tendon, push a button, and hoist a steer for skinning. For the archaeological context, improvements in cattle seen with the influx of the Romans also led to changes in butchery. On the one hand, cleavers seem to have become the mainstay of the craft on urban and military sites, rather than small general-purpose knives. There is also evidence that the carcass may have been suspended (Hambleton & Maltby 2009). However, it is probable that butchery of large domesticates was undertaken by a group of individuals, particularly in assisting with skinning.

Expectation 3: Sequence reveals behaviour.

A butcher will make decisions regarding how best to remove the limbs from an animal, for example, based on a range of factors: the presence of helpers for cooperative butchery, or the available implements. However, if the leg needs to be removed quickly, and the implements are present to chop through the bone, then the butcher might employ a specific technique to do this: slicing into the muscle of the thigh and chopping the ball-and-socket joint at the head of the femur. In this case, the rationale of speed acts as the catalyst for the desired outcome. Without this driver, there would be little requirement for cleavers or the technique. The motivation for both the technique and tool might be progressive: intensification in demand over time.

Alternatively, the driver could be more immediate: an isolated short-term overabundance of animals that need to be processed quickly, for example by an itinerant butcher. These catalysts create a principle that foments the actions, the behaviour, required to complete the task: a need to disarticulate the carcass rapidly. Components of local identity, ritual and sacrificial practice, or cuisine and nutritional requirements could all potentially serve as drivers for distinctive butchery. For example, is there a need to maximise the energy yield from the carcass as a whole? In this case, one might expect to see meat removal marks on all areas of the carcass, not only limbs, as well as split long bone shafts and fragmentation of elements with low meat yield, such as phalanges.

Other expectations that can help describe human behaviours relate to the sequence of butchery as mediated through morphology: when certain carcass parts are disarticulated. This includes the probability that removal of peripheral elements, such as the head and lower limb extremities, will occur at an earlier stage during the processing of the carcass. This principle shares similarities with a phenomenon that has been widely discussed within the archaeological literature, 'the schlepp effect': the distance that needs to be travelled back to a camp, from the site of a kill, will dictate the portions of the carcass that are removed (Daly 1969: 149). The principle has been gleaned from work with modern-day hunter-gatherer communities and is generally applied to prehistoric archaeological contexts (Lyman 1987). For the modern commercial context, motorisation and refrigeration have long since removed limitations on transportation. However, the schlepp effect actually encapsulates a more important aspect of cognition: what drives individuation during the process of butchery? The removal of specific elements will occur in an order that facilitates subsequent butchery. For example, by detaching the head, there is a reduction in the overall length of the carcass, thus it will not need to be hoisted as far from the ground for further disarticulation (assuming the animal is being processed while suspended). Such precision during butchery has become essential due to the volume of carcasses that require processing.

Some of these topics will be discussed further in later chapters. The aim here is to stress the ways in which contemporary practice and expectations have influenced my inferential framework (Chapter 9), with a view to applying a practical working knowledge of this complex food industry to the study of archaeological butchery. Finally, the examples above are not presented to suggest 'direct correlation'; rather, they illustrate the varied ways in which the wider view of modern butchery can resonate with and help us interpret the archaeological context.

CONCLUSIONS

The power of modern analogy lies in drawing our attention not only to technical and technological aspects of butchery but also to the details of meat retail. These may not always be applicable to the past, but they bring features to our attention that we might be hard pressed to understand from the archaeological evidence alone. Some examples of modern practice serve within a comparative dimension, demonstrating parallels between the paraphernalia and products associated with meat selling (Mac Mahon 2005: 70–88). The modern context brings the breadth of practical considerations that need to be situated alongside the social dimension. Modern analogy can also help to clarify the know-how of the craftsperson. Butchers do not view a carcass from the perspective of the animal's skeletal morphology; they assess it from the musculature, tendons, sinew, and fat – the soft resources. The skeleton needs to be negotiated or mitigated, in order to gain various products, which can include the bone itself. The degree to which the skeleton has to be manipulated will be intimately tied to the technologies that the butcher has at his or her disposal. As O'Connor observes from the site of Coppergate in Viking York, 'the atlas served as part of the head for the purposes of butchery' (O'Connor 1989b: 154). Recognising this distinction more accurately reflects how the butcher conceived of the animal. In turn, this type of reasoning assists in our own interpretations of how the carcass was divided and processed. Achieving this improved interpretation depends on establishing a distinction between 'butchery' and 'carcass' units: the latter being represented and based on natural morphology and body design, the former on how the animal has been processed. This could serve as an axis for assessing different approaches to dismemberment and provide an expected baseline. Thus, this last point, and the one made above vis-à-vis efficiency within carcass processing, offer opportunities to develop predictive models that could have utility for developing more rigorous comparisons across sites (Seetah 2008).

As archaeologists, the material we study has passed through a series of stages, each of which is driven by an increased level of complexity in economic or social significance. This emphasises the usefulness of skins, hides, and horn, along with bone for working. However, the economic viewpoint needs to be tempered with a deeper appreciation for subsistence. This involves recognising and, if necessary, removing our modern cultural attitudes in the way we view animals and meat. A more astute reading of the archaeological record requires the analyst to situate, if possible, the nutrition (as well as economic value) available from skin fat, marrow, bone epiphyses, intestines, internal organs, the brain, etc., while bearing in mind that 'nutritional' value is itself, in part at least, a cultural construct. A complete evaluation will not be possible from the archaeology, but some of these details may well be revealed through a more thorough reading of the cut marks. It may also be useful to view carcass processing from the perspective of greater efficiency in terms of exploited resources – to consider contexts where everything is taken from a carcass because it is absolutely necessary for survival, where eating takes place out of necessity rather than desire.

For the archaeologist, the fact that the material we study is almost universally fragmented means that we have a particular approach to the data, and one that stands in contrast to the ancient butcher. Modern analogy could help to situate the seminal place of 'the carcass', the value of non-food elements of the carcass, and the context of conspicuous disposal. In short, analogy helps to position a wider conceptualisation of the animal and its parts and, indeed, the living organisms. These cues are further discussed in Chapter 8, using analogy to better situate gross morphology, independently assessing soft and hard tissues, behaviour, husbandry, etc., adding some of the details that are absent from the archaeological record.

CHAPTER SIX

THE MATERIALITY OF BUTCHERY

Here animals met their fate at the hands of workers and machinery, creating a vast 'disassembly'
line that ended not just the lives of pigs but the age-old relationship between meat and mankind.
(Pacyga 2015: 1, describing Chicago's Union Stock Yard)

INTRODUCTION

Performance, gestures and bodily movement, and enskilment all have an
inherent materiality (Matthews 2005). The 'techniques of the body' depend
on learning and using specific objects, whether tools, such as spades, or shoes
(Mauss 1973). Continuing the focused discussion on technology, this chapter
now turns to the material culture associated with butchery. There are typically
many objects associated with the craft – from butchers' blocks to whetstones,
and meat hooks to scales. However, these are rarely recovered from the
archaeological record (e.g., butchers' blocks) or are not solely part of the
butchers' tool-kit (e.g., whetstones, which are used by a range of artisans).
For these reasons I focus specifically on knives, the most versatile and commonly recovered metal implement from archaeological excavations (Tylecote
1987: 262). Viewing the functional properties of knives through a modern lens
can reveal details that are not generally covered in the archaeological literature
describing these implements (Bouchnik 2016: 306; Greenfield et al. 2016: 89).[1]
In this way, archaeology benefits from a more complete assessment of the
tools' practical capabilities, which can be integrated with evidence describing

typology and provenance. Pacyga's quote exemplifies that nature of 'deconstruction' introduced in Chapter 5. This chapter examines the knife, its features and characteristics, and how those make this object singularly suitable for the task of butchery. It is through the knife that the enactment – the performance – of butchery is expressed.

Furthermore, returning again to the model of butchery as an 'ecosystem', knives operate as an essential component of this broader network. Tools offer a material lens on the complexity of butchery as a production process: food, humans, animals, and raw materials like iron come into interaction around the edge of a knife and are guided by the agency of farmers, blacksmiths, and the craftspeople involved in animal body-part trades. From this vantage point, we can appreciate how a vast array of drivers, from agriculture to the military to food practices, influences the production and use of this 'expensive' object. The chapter moves through different scales and facets of the butchers' craft, using objects to better understand agency, behaviour, and mindset.

IMPLEMENT SPECIALISATION: FIRST STEPS ON THE PATH TO MECHANISATION

The transition to metal from lithics is evidence of profound developments related to modes of resource procurement, and production processes required to create the tool (Tylecote 1987, 1992; Harding 2000; Chapman 2003; Kristiansen & Larson 2005). These go hand in hand with a major intensification in pastoralism, animal husbandry, and agriculture, particularly in the Levant and Near East, from the end of the Neolithic period (Yener 2000). While the Near East has been an important centre for research on this transition in material exploitation (Maddin et al. 1999; Levy 2007), many other regions have also provided valuable evidence on the details of adoption of this new technology, for example, the Urals (Linduff 2004) and the steppe region (Frachetti 2002: 161–70; Hanks 2010). One feature that seems evident across the Near East, Central Europe, and Britain is that the uptake of metal for butchery was not immediate (Olsen 1989; Greenfield 2013b; Muhly 2013). Greenfield's research in the Levant has revealed that lithics continued to be the favoured raw material for carcass processing, despite the ready availability of 'better' metal tools (Greenfield 2013a, 2016: 273). This is one of the more revealing details of the transition to metal. This delayed adoption could rest on the fact that the knowledge of butchery was intrinsically tied to another material, lithics. The shift to metal may then have occurred as a consequence of external drivers, perhaps as a result of intensification in meat consumption as a consequence of centralised settlements, although it is possible that this transition may have occurred much later.

Regardless of the precise timing of the adoption and widespread utilisation of metal tools, most zooarcheologists (with exceptions, see Olsen 1989; Greenfield 2016) have paid little attention to the connection between changing trends in the type of material used for butchery tools as the technology itself develops. This is important in terms of our ability to understand and situate the larger drivers: is the transition universal and predictable? What are the mechanisms for dispersal and adoption of new materials and techniques of production? Are changes in the materials used for butchery tools driven internally, for example by the need for implements with a keener cutting edge, or externally, as part of the comprehensive advance of metallurgy? This latter point is best illustrated through the example of innovations made in weapons-making technologies that eventually filter into quotidian practice (Tylecote 1987: 262). As examples from modern non-Westernised cultures indicate (Seetah in prep b), it is probable that butchering was not a major force for driving innovations in metal technology until the craft itself developed into a more standardised and specialised activity. This is also borne out in the archaeological record, where early metal tools from Iron Age Britain, for example, do not demonstrate the same degree of variability when compared with later periods (Manning 1979: 118).

Turning our gaze onto the tools themselves, a significant interpretive gap emerges when we consider that implements recovered from a given site or period are rarely analysed in tandem with the cut mark data. Do the tools recovered correlate with the types of butchery observed? Only rarely are we likely to recover the tools that were used to create the cut marks. The goal is less to posit direct correlations but more to generalise on the activity taking place at the settlement level, regionally, or diachronically. The variety of tools supplements our ability to identify general-purpose versus specialist implements and makes it possible to ask how the transition from one to the other took place over time, and why, and to ask who was using the tools. These questions may only be addressed in general terms. However, this allows us to corroborate and situate our cut mark data within its wider socioeconomic setting, taking into account gender, social status, cost of production, and the acquisition of raw materials.

SPECIALISATION OF TOOLS FOR BUTCHERY

The remainder of this chapter examines specific aspects of the technology and techniques that underpin the use of knives during butchery. The archaeo-historic aspects cover the Iron Age to medieval period in broad chronological terms; specificity is difficult as few metal objects are accurately dated. Tools are created for cutting animal bodies, and therefore, by extension, butchery is both a driver of and driven by implement production: what are the ways in which the evolution of blade technology directly connects to butchery? In short, blade technology and function should highlight the individuality of specific stages within the dismemberment process and the value of tools for discrete cutting tasks.

The Material Basis of Knife Manufacture

Production processes have obvious bearing on the composition of the tools produced, which in turn influences function. Our knowledge of iron artefact production comes from the waste following smelting, examination of the finished product, and, to a lesser extent, literary and iconographic sources (Cowgill et al. 2001: 32). Of these, investigation of the excavated implement is the most important for the purposes of this book. It is possible, for example, to determine how long the blade spent in the forge, or the approximate temperature of the furnace, based on the amount of carbon diffusion. Details of manufacture are revealed using radiography; two-thirds of makers' marks would not be visible without X-raying (Cowgill et al. 2001: 74).

For Britain, by the time of the Roman conquest in AD 43, the production and processing of iron was a well-established activity (Himsworth 1953: 36; Tylecote 1992: 54). Three main types of iron were available to Iron Age smiths: cast, wrought, and steel. Cast iron is the least pure form, with a high carbon content. The inclusion of carbon in iron makes it harder. However, at the carbon levels that are found in cast iron, the material is made brittle. Cast iron is so-named as it can be melted and poured – cast – due to this high level of carbon. Wrought iron has few impurities, usually making up no more than 1 per cent of the overall composition, and is created from a smelted bloom of iron and slag. The iron is welded together and the slag expelled through heating and subsequent working (Tylecote 1987: 248). The composition of steel places it in between wrought and cast iron. It is harder than wrought iron, but not as brittle or prone to breakage as cast iron. There are a number of ways of making steel from wrought iron. The underlying principle is to increase the amount of carbon to between 0.01 per cent and 1.7 per cent. The higher the carbon content, the greater the effect on the raw steel, making it stronger and more durable (Woods & Woods 2000: 30).

In terms of the chronology of when these metals were used for edged implements, the intense furnace temperatures required to achieve a high carbon content meant that cast iron was not used to any great extent until the fifteenth century. Furthermore, the fact that it was brittle meant it was not used extensively, evidently not for knives or other implements (Manning 1979: 118). Wrought iron has been the most commonly used metal since the Iron Age (Manning 1979: 117), up to the modern period, following which methods to produce inexpensive steel led to the latter material superseding iron in the commercial manufacture of metal goods (Tylecote 1992: 167). From the Romano-British period onward, the carbon content of iron was in the region of 0.6 per cent, resulting in a hard, but reasonably tempered, metal. Finally, the production of steel during historic periods depended on prolonged heat treatment at a consistent and high temperature, often with carbon-rich materials such as horn, bone, and leather packed closely with the wrought iron in fireproof boxes. This resulted in carburisation of the surface and a diffusion of carbon into the underlying structure, effectively creating steel. Steel has been estimated to cost approximately four times as much as iron (Tylecote 1987: 278) and was therefore considered a premium raw material. Aside from expense, it was difficult to produce early steel to a consistent quality because of the prolonged and stable heat needed during the smelting process. For these reasons, steel was expensive, and good quality steel, with the right balance of hardness and flexibility, was highly prized.

Thus, the majority of knives during both the Romano-British and medieval periods were produced from wrought iron; however, despite serviceable durability, this type of iron did not retain a long-lasting, sharp cutting edge in the same way as steel could (Ward 2008: 32). The addition of steel into the cutting edge, or 'steeling', is indicative of intentional inclusion for the express purpose of increasing hardness at that location (Tylecote 1987: 259; Cowgill et al. 2001: 8). Nonetheless, the cost, coupled with the fact that welding steel to iron could potentially result in 'de-carburisation' of the steel (making it softer), meant that the material was used sparingly.

Knife Fabrication, Components, and Design

The technology for 'steeling' existed from the Iron Age and was certainly well established by the Romano-British period (Manning 1979: 118). However, this technique probably was not combined into knife technology in a consistent manner until the medieval period, when considerable improvements emerged in technical standards compared with the Roman period (Tylecote 1986: 197). During the Romano-British period the inclusion of steel into metal implements seems to have been reserved for swords and other weapons, and less commonly used for knives (Tylecote 1987: 262). In many instances, knives were not even carburised by heating in order to harden the wrought iron itself (Tylecote 1986: 177), although 'work-hardening', a form of cold working the metal to increase hardness, could have functioned to maximise durability of the cutting edge (Tylecote 1986: 174, 199). As the following text elaborates, this is crucial for functionality, particularly within a butchery context. The inclusion of steel thus becomes indicative of particular requirements in the knife itself. These functional properties – such as a sharp and durable cutting edge – would have utility for a range of tasks. However, as few knives in the Roman period

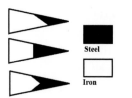

FIG. 6.1 Schematic illustration of three methods to combine steel and iron in knives.

show evidence for the inclusion of steel this would suggest that those implements with steeled edges were likely created for specific activities.

'Steeling' of the Cutting Edge

During the Romano-British and medieval periods, steel could be combined into a knife in a number of ways (Tylecote 1986: 174; Cowgill et al. 2001: 10), after which the implement was quenched (rapidly cooled in water) to increase the hardness of the steel component. The quenching process also resulted in the knife becoming brittle. To counter this, the implement was 'tempered' prior to final shaping by gentle heating, and then allowed to slowly cool (McDonnell 1989; Cowgill et al. 2001: 8). Figure 6.1 illustrates techniques for combining steel into an iron knife, highlighting the variety of methods that were in use (see Cowgill et al. 2001: 10, fig. 5, for a more complete representation). Cost played an important role in determining which technique was used; however, experimentation seems to have been an additional catalyst. These differing methods of combining metals may reflect the needs of specific crafts, professions, or trades. Steeling the edge may be a key distinction in identifying specific butchery knives from the archaeological record.

The Handle

The manner in which a handle is attached to a blade and the provisions made during manufacture of the blade for subsequent handle attachment have a number of implications for the functional capabilities of the tool. Three main types of tang (the portion of the blade that attaches to the handle) were used during the Roman and medieval periods: the whittle tang, the scale (also known as a slab) tang, and through tang. The first and second of these are discussed in greater detail here. The third type is similar to the whittle tang, with the difference that it is attached along the full length of the handle, rather than only part.

A whittle tang knife can be created from a single piece of wrought iron and is the simplest and least expensive means of manufacturing a handle (see Fig. 6.2). It involves drawing out a section of the blade into a point. This is usually completed prior to the fabrication of the blade itself. The drawn-out tang is used as a gripping point for the tongs to allow the rest of the blade to be forged. Once the knife has been completed the tang can be heated and forced into the handle, resulting in a seared connection between the blade and the handle. This type of arrangement is best suited for smaller, lighter knives.

Function also dictates that this type of handle is best suited for smaller knives. The attachment between the blade and the handle is a relatively weak one. Therefore it is prone to movement. This is further affected by shrinkage

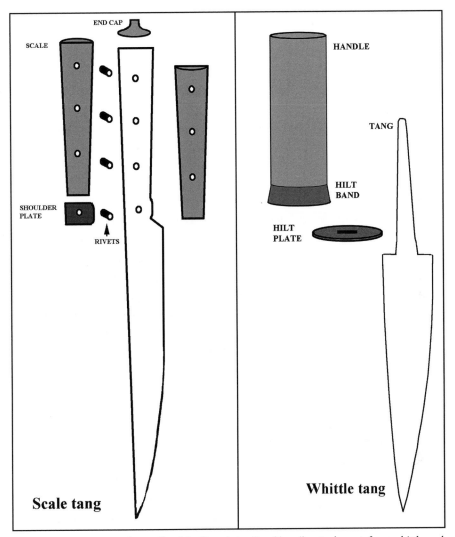

FIG. 6.2 Components of a medieval knife and details of handle attachment for a whittle and scale tang.

in the material used to make the handle, or during use. Thus, with a longer knife, there is a greater likelihood that the blade will become detached from the handle. For butchery, this type of handle is best suited to compact blades, used for skinning and other tasks where relatively little force will be generated at the handle. There is also the risk that the blade will snap at the tang if too much force is exerted, especially with thin knives. Thin blades are ideal for retaining sharpness, which is an essential functional characteristic; the smith can also minimise the raw materials used during manufacture. There are thus a number of incentives for producing a thin blade, even if this renders the implement more prone to breakage. This illustrates the balance that needs to be achieved in order to produce functionally durable knives.

Where greater durability is required, a scale tang is used (Fig. 6.2). More material is required for the production of a scale tang knife. However, it will be less prone to breakage, is unlikely to detach from the handle, and can be forged into a greater variety of tool types, making it a more versatile design. In this arrangement the tang is a more substantial component of the knife. The handle itself is attached with rivets forming a stronger bond with the blade. In effect, the handle is included for comfort and/or decoration; the knife could still be employed in its full functional range without the handle.[2] With a whittle tang, the handle is an essential component of the knife's functionality.

A scale tang (or variation thereof) is employed for knives that are used for more forceful cutting or those with a larger blade. A key component of the scale tang is that the handle is riveted to the knife. Thus, this type of tang can be longer compared with a whittle tang, and thicker. Increasing the thickness results in increased durability: it is more difficult to break the knife. These factors are crucial for the production of cleavers. While the blade is of central importance, a cleaver could not be used for chopping activity without a strong tang and handle. Variations on this theme do exist; for example, the cleaver replicated from the Romano-British period and discussed in Chapter 10 uses the 'wrap around' method. In this instance the tang is formed to fit around a wooden handle and riveted for added durability. The development of this arrangement of tang underpins the unique design of cleavers. Contemporary butchers reap the benefits of this innovation to the present day.

The Blade

The cutting edge of a knife, its blade, is the most important part of the implement. Variations in the shape and size of the blade affect how the implement will be used for butchery.

The sharpness of the blade's edge is key to functionality. Two main factors affect sharpness: first, the hardness of the raw material, as discussed above, and second, the thickness of the blade. It is easier to maintain a keen edge on a thin blade. Making a thin blade from a material that is relatively hard and brittle, as in the case of iron, must have been a challenge to early smiths. The only way to achieve such a blade during the Romano-British and medieval periods would have been to hammer the forged blank blade before final shaping. In the modern setting the blade is ground down using abrasive materials. Thinning the blade through forging would not have been a precise process, and early archaeological knives reflect this in that they are generally thicker, and less hard (Blakewell & McDonnell 2007), than modern implements.[3]

The need for a sharp blade reveals some fundamental aspects of carcass processing. It might be assumed that force plays a significant role in butchery. However, when using knives, or even cleavers, strong actions are to be avoided. Primarily, this is to minimise the potential for cutting oneself should the knife slip. Moreover, particularly when chopping against a butcher's block, momentum, not force, is utilised. Thus, in all circumstances the butcher will favour a sharp blade, as this will reduce the amount of effort needed for cutting into the meat or joint. Anecdotally, this is reinforced through an often-used axiom in the profession: 'the only dangerous knife is one with a dull edge'.

While a butcher will want as sharp an edge as possible, the shape of the blade is open to variability. As with all components of the knife, shape is closely dependent on the materials available. Large cutting implements for domestic or commercial use are relatively rare within the archaeological record, particularly from the Romano-British and preceding periods (Manning 1976, 1979: 118–19). This might be a factor of cost or availability of raw materials. However, it is possible that this has to do with the requirements of the meat processors themselves and their interaction with the blacksmith. Variation in shape provides evidence of different functions.

The clearest example of this is seen in the differences between cleavers and general-purpose blades. Cleavers are employed to chop into bone, whereas knives are used to cut around bone or into meat. In some instances large blades have been erroneously categorised as cleavers (Collingwood & Richmond 1969: 319, see plate XXI e).

The key difference between these two tool types is the handle. The cleaver, for reasons discussed above, will invariably have a scale tang, while a blade is more likely to have a whittle tang. The shape of the blade will also be different. The cleaver will have greater depth from the cutting edge to the top of the blade. It will usually have a region of the blade, close to the back of the tool, which is relatively flat to allow for chopping. Modern cleavers are generally flat along the length of the blade. This provides for a greater area of contact with the carcass during chopping. Modern cleavers are effectively only used as a chopping tool. In contrast, both Romano-British and medieval cleavers tended to have a certain degree of curvature to the front portion of the blade, particularly noticeable in the Romano-British cleaver. Shaping a blade to incorporating a degree of curvature facilitates slicing as well as chopping. However, variations exist and are driven by regional factors as well as butchery style; for example, contemporary cleavers in some parts of continental Europe still retain a marked curvature (Fig. 6.3).

Other functional aspects of shape relate to the degree to which the tip of the blade is pointed. Skinning and boning knives have an acute tip. This allows the knife to be used to puncture the skin during the initial stages of skinning. Furthermore, if the tip of the blade is relatively long, a greater portion of the blade can be kept against the bone during meat removal (see Chapter 9 for discussion on using these characteristics to assess cut marks). The profile of the cutting edge also influences function. The cutting edge of a cleaver, which usually has a relatively thick blade, cannot be honed to the same keenness as a

FIG. 6.3 Modern cleaver with pronounced curved cutting edge, from Ronda, Spain, a former Roman enclave. (From the author's collection.)

thinner knife.[4] A general-purpose knife, on the other hand, is suitably thin throughout the whole depth of the blade to permit rehoning by hand throughout much of the life of the tool, although this does not negate the need for occasional re-edging.

The final parameter of interest is size. Increasing the size of an edged implement would have facilitated the butchery of larger animals. However, this is not due to a straightforward correlation between blade and animal size, but rather between blade weight and bone density. This is specific to the type of butchery we see from the Romano-British period, which was heavily reliant on chopping activity. By increasing the size and weight of the blade, the durability of the implement is also increased, which is essential when the tool is to be used for prolonged and consistent chopping of dense bone. Weight, as a separate but related component of overall size, plays an import-ant role in how a cleaver is used. Butchery that depends on chopping (as opposed to knife cutting) relies on the weight of the implement in order to provide momentum. This reduces the force the butcher needs to apply and facilitates butchery.

The above has drawn on modern analogy, replication studies undertaken as part of the research for this book (Seetah in prep c), evidence from archaeometallurgy, and modern texts on knives to substantiate the func-tional properties of archaeological tools. The summary identifies subtle differences in knife manufacture and design and the implications this has on modes of use. It is likely that specialist butchery implements did exist at least from the Romano-British period. However, recognising these from the archaeological record depends on the identification of variations in knife technology that show a correlation with changes in butchery tech-niques. Perhaps more important is considering how to integrate the arch-aeological evidence in meaningful ways to better connect metal finds with the cut mark evidence.

ARCHAEOLOGICAL RELEVANCE

As we have the physical remains of implements in the archaeological record, we can have a clear understanding of the material properties and fabrication processes of knives. Given the ubiquity of metal knives in the modern setting, it is relatively easy for us to acknowledge their place within the butchery profession and as part of the craft. Some tools have a variety of uses; others have a limited repertoire and are employed for specific tasks. In the absence of more specialised knives, certain tools may have to be repurposed. Thus, knives serve as the direct interface between the thought process of the butcher and the outcome in the form of a butchered carcass. How do we transition from looking at the provenance and typology of metal finds to situating this inherent behavioural complexity?

Artefact studies are an obvious strength of archaeology. By identifying and examining the drivers for tool production, we may be able to better understand why the implements were made in the manner we observe, and to interpret details of their use. The chapter has illustrated the ways in which knife construction influences function, and how changes in fabrication over time can exemplify broader socioeconomic or ideological dimensions. Kosher practices, specifically religious slaughter termed shehitah, and the physical characteristics of the knives used within this context, provide an important example of the relationship between tools and social context. Given the fact that religious dietary regulations can lead to discrete but lucrative markets, it should be evident that this relationship incorporates economic dimensions also (see Fishkoff 2010 for growth in kosher produce in the United States, and Lytton 2013 for corruption during inspection as a consequence of growth).

In Jewish practice, only certain animals can be considered kosher, and only meat that has undergone shehitah is fit for consumption (Blech 2009, 137–40). The shochet, a specialist butcher/slaughterer, undergoes a long apprenticeship that includes induction in both the correct ways to butcher and rabbinical laws regarding consumption (Lipschutz 1988: 21). Shechita is remarkably precise. The stroke to kill, the physical act to extinguish life, is arguably one of the most prescribed 'cuts' any butcher performs. The stroke – here describing the path of the cut – must be continuous, without any hesitation, pressure, or tearing. The stroke can be made with both a forward and backward action, but pressure must be constant. The cut itself must be made in a specific area toward the middle of the neck, severing the trachea, oesophagus, carotid, and jugular (Blech 2009, 139), but without making any contact with the bones of the vertebrae (Greenfield & Bouchnik 2011). The cut must be made without any covering (*chaladah*), e.g., the wool around a sheep's neck, of the *chalaf* (*sakin* or *hallaf*: the knife used for slaughter).

Given these exacting requirements of the cut, one can appreciate how vital the *chalaf* is to shechita and Judaism. The blade is fabricated to specific proportions and has to have certain characteristics. It must be at least twice the length of the width of the animal's neck but cannot be heavy relative to its overall dimensions as this may cause undue pressure. The blade cannot end in a point, as this may inadvertently slip under the skin, and therefore become 'covered', rendering the animal unfit. A serrated edge is unacceptable, as this would cause tearing. The surface of the blade must be smooth to facilitate inspection (Verṭhaim 1992: 302). The blade has to be razor-sharp, without any imperfections: 'the cut must be made solely by the sharpness of the blade. Even placing a finger on the blade renders the *schechita* unfit' (Lipschutz 1988: 20). When slaughtering mammals, the *chalaf* is inspected before and after each animal (Blech 2009: 140).

The social and ideological importance of material and fabrication is exemplified in the use of 'whetted knives' by Hasidic Jews, a religion sect founded in eighteenth-century Ukraine that spread throughout Eastern Europe. Whetted knives were made of molten iron and whetted (sharpened/honed), making the edge sharp and blade smooth to the required standards. Hasidic Jews used such tools in response to the fact that blades at that time, when made from forged iron, could not apparently be honed to be adequately sharp and smooth. The use of this type of blade, which could be polished to a mirror finish, caused considerable controversy, effectively rendering animals killed by Hasidic Jews to be considered unfit for consumption by other Jewish groups. Hasidism was excommunicated in 1772, due in part to the use of whetted knives for shechita (Verṭhaim 1992: 302–15).

Work by Greenfield and Bouchnik (2011) describes how features of religious ideology surrounding shechita can be observed in the archaeological record. Their research incorporated archaeological evidence of slaughtering practice as depicted in iconography and from the bone assemblages and historical evidence gleaned from religious texts concerning the activities of shechita and from historic and contemporary accounts denoting the physical properties and functionality of the *chalaf*. The results suggested consensus between archaeological butchery practices from sites in Jerusalem and religious texts. Similarly, Jewish communities have been identified from animal remains recovered from medieval Catalonia (Valenzuela-Lamas et al. 2014).

The case of shechita and similar prescriptions for methods of slaughter in Islam, for example, illustrate the connection between the techniques of butchery, the technology of knife production, and the wider socioeconomic context. For Jews in the United States, the hindquarters of mammals effectively became non-kosher and were sold to non-Jews, because no one had the skills for *nikur*, the removal of veins and arteries, in those early immigrant populations (Lipschutz 1988: 29; Blech 2009: 146). The practice of selling

FIG. 6.4 A *chalaf*, used for kosher slaughter, in this case for fowl; note the squared end. (From the author's collection.)

hindquarters has a long tradition where communities of Jews and non-Jews lived together, but it is revealing in the US case given the specific cause, in this case, a lack of butchers skilled in this aspect of meat preparation. Beyond the context of trade and exchange, the relationship between slaughterer and blacksmith for production of such an exacting blade as the *chalaf*, must also have been a singular one. Given the requirements for a blade of specific dimensions relative to the animal being slaughtered, one can appreciate that numerous knives formed part of the tool-kit of the shochet (Fig. 6.4). Shechita and the attendant materiality highlight important dimensions between craft, human agents, and the animals that formed the backbone, for butchers, of their trade. Furthermore, this case illustrates the importance of social paradigms within the process of commodification of animal bodies and their parts.

As a way to add complexity, in concluding this section I would like to offer counterpoints to the case presented above, describing the significance of the *chalaf*, and a number of more general notions that the last two chapters have emphasised. Ideas of 'efficiency' and 'specialisation' within a butchery framework have run implicitly through this discussion. These concepts both permeate the archaeological view of butchery (Machin et al. 2005) and exemplify the modern, mechanised industry. In one sense, efficiency can be represented, even quantified, by the time spent on the tasks of butchery that have been discussed in Chapter 5. Equally, specialisation, as described in this chapter, is characterised by the fabric and fabrication of precise cutting tools.

In 2010, I undertook my first ethnographic study in Kenya, which formed the basis for subsequent work that I have described in other parts of this book. The research in 2010 was undertaken in collaboration with a group of Maasai from the Kuku Game Ranch, Oliotokitok, Kenya, approximately four hours' drive south of Nairobi. The aim of this study was to gain insight into Maasai cosmology as it related to meat consumption and centred on the slaughter, butchery, preparation, sharing, and consumption of a young male goat (Seetah in prep b). The example presented by this case study illustrated all the technical sophistication and social complexity that characterise 'butchery' as craft, practice, and performance but provides a view in which certain economic dimensions are absent.

Daniel Mabuvve, a colleague who undertook the slaughter and processing during the field experience I describe, was obviously a skilled butcher.

FIG. 6.5 Butchery of a goat using a traditional Maasai approach, the two knives that were used for processing can be seen resting on the goat's neck area.

However, little preparation went into selecting a suitable site on which the butchery was to be performed. An open space was chosen that was reasonably close to the kitchen area, and a large piece of flattened-out cardboard was placed on the ground. Daniel chose two knives. On questioning, it was clear that this was an ad hoc choice of tools, with few requirements placed on the implements themselves. The knives he chose happened to be readily available from the kitchen. They consisted of a lightweight cleaver, similar in design to a Western, commercial cleaver, but fundamentally different. This tool did not have a heavy, thick blade; the handle was plastic; and the tang was not socketed but forced into the plastic. The small cleaver, while useful for light chopping, would not have been suitable for fracturing dense bone. The other tool was a thin-bladed filleting knife, highly practical for skinning and similar tasks but, once again, with a plastic handle and therefore less durable. Clearly, the knives held relatively little significance (Fig. 6.5).

There was no pretext at ritual; the animal was simply being prepared for dinner! Daniel's knowledge of the animal's anatomy, the set pattern of butchery he followed, and his control of the knives as he guided them to achieve specific outcomes all resonated with my own experiences of carcass processing. However, Daniel showed no urgency. Time simply was not one of his concerns, nor did he appear to be driven by notions of efficiency. Daniel was guided by cultural norms, and his approach to butchery emphasised the importance of social attitudes to cuisine and sharing, and how these motivated practice.

For the archaeological context, the examples of shechita and Daniel's approach to butchery signal the importance of maintaining a balanced view, and why the evidence from cut marks can be more effectively exploited if nested within technical, social, and economic frameworks. In short, butchery is contemplated 'as part' rather than 'apart' from the wider archaeological context.

CONCLUSIONS

A recognisable phenomenon exists: cutting tools are improved to facilitate carcass processing. The craftsmanship to create tools specifically for processing flesh has a long-standing precedent. It might arguably be said to have reached a zenith in contemporary Japan, developing from the tradition of sword making (Kapp et al. 1987); '*katsu-sju-ho-ju*' – cut first, then simmer – illustrates the axiom that cutting is an essential component of cuisine (Nozaki 2009: 10). Knives created for precise cutting of flesh, particularly fish, are recognised globally as some of the most sophisticated and exacting implements created for butchery. Such tools, alongside those created for chefs and butchers through refined craftsmanship, illustrate that all knives are not created equally (De Riaz 1978).

Transitions in butchery, e.g., using metal instead of stone tools, or large versus small blades, deputise for shifts at an economic, social, technical, or settlement level. The production of meat follows the same principles as other goods: resources are located and extracted, whether by hunting or raising animals, and a commodity is created through butchery and cooking; finally, there is consumption. The difference in this scenario compared with other products rests on the context of intentionality and on patterns of mobility. At one end we have an animate – sentient – resource and, through butchery, a product at the other end. Intentionality of the human agent is key, and this is exemplified through the use of tools. Thus, the context of those objects that are used during the transformation of flesh to meat must form a more cogent part of our analysis of butchery.

The composition, production, and functionality of the tools have implications for a range of other features that revolve around butchery and the butcher. Referring back to the techniques for butchery discussed in Chapter 5, the range of cutting practice for which knives are used, from chopping to point and blade insertions, etc., effectively illustrates an evolution in butchery – from simple slicing marks – that was likely initiated with increased demand and made possible with new and different tools.

Analogy can be applied in more nuanced ways, by thinking phenomenologically. Larger and more durable tools are *perceptually* more functional. There is less fear of breaking the tool; therefore, it is used in a way that is more forceful. They can also be used more often simply because they are more durable. Once *mindset* has shifted to a new mode of butchery, new techniques can also develop. The conceptualisation of the tool also changes. When knives become more durable, they become more necessary to the butchery process because they replace some of the need to manipulate the carcass and mitigate the need for forceful 'breakage'. In effect, a better knife will end up being used for a greater proportion of the butchery process, and this increases the significance of the tool.

As a case in point, it is interesting to consider who might have used specialist knives, i.e. cleavers or steel-edged blades. Butchers who could afford such tools must have been important patrons of the blade smith. Taking into account the sheer quantity of animal bone recovered from large, historic urban sites leads us to reflect on who else would need knives that could consistently chop bone, or at least, why else would such knives be created if not for butchery, whether commercially or in the homestead? We have largely missed the significance of metal implements as a line of evidence for interpreting butchery and butchering, which stands in sharp contrast to the emphasis on tools that has been the hallmark of research into lithics.

The preceding three chapters illustrate the potential of modern analogy to deepen our understanding of butchery. This can be by means of a more direct correlation between different technical or technological practices, or through a phenomenological perspective, providing useful points of insight for interpretative archaeology. In Part II, the book uses ideas developed in Chapters 1–6, revolving around technology and socioeconomic salience, as points of departure for discussing how we might better assess the archaeological context.

PART II

THE PRACTICE OF BUTCHERY IN
ARCHAEO-HISTORIC SOCIETIES

CHAPTER SEVEN

STUDYING CUT MARKS IN HISTORIC
ARCHAEOLOGICAL CONTEXTS

INTRODUCTION

Within archaeological discourses on food production and consumption, the question of agency has moved from implicit to explicit. Early references to hand-axes as 'cultural fossils and the product of the human mind and human craftsmanship', or observations that archaeologists only have access to the 'cutlery . . . of a society' (Daniels 1963: 30, 132), have given way to more concrete integration of 'fauna as food' into our assessment of the past (Bettinger et al. 2015). The place of animals and their remains in archaeological analysis rose in significance with the advance of processual archaeology. Animal bone came to represent a crucial dataset, in part because these finds could be quantified in numerous ways, but also due to the economic significance of fauna to past societies (see Binford 2001). For post-processualists, animals continued to be important, but in contrast to processualists, it was their role as 'symbol' that was particularly relevant (see Flannery & Marcus 1998: 45, table 1; Jackson 2014: 107). Increasingly, however, scholars have called into question the utility of such dichotomies, between economic and social, or subsistence and symbolic. Using a number of cases drawn from a wide range of geographic regions, Russell, for example, illustrates the social importance of slaughtering and butchery practice and consumption during ritual ceremonies (2012: 88–127).

For butchery specifically, early work by Lartet (1860) led to the genesis of cut mark analysis, although White (1952, 1953, 1954, 1955) is credited with developing the subject along the lines followed today. The study of butchery was central to methodological developments in zooarchaeology (Guilday & Parmalee 1962; Yellen 1977; Brain 1981) and influenced archaeological thought through middle-range theory (Binford 1978).

This early work formed the basis for a large body of literature on cut marks (e.g., Aird 1985; Lyman 1987: 249–337; 1994: 294–352; 2005; Reitz & Wing 2000: 128–31; Domìnguez-Rodrigo et al. 2005, 2008; Dewbury & Russell 2007). In particular, recent research has drawn attention to the importance of cut marks and butchering behaviours within the context of technological development, subsistence strategies, and sociocultural idioms (Mowat 1970; Mooketsi 2001; Domìnguez-Rodrigo 2005). Butchery traces have been studied from experimental (Jones 1980; Lupo 1994, 2002, 2006; Braun et al. 2008) and methodological perspectives (Hill 1979; Aird 1985; Blumenschine et al. 1996; Lignereux & Peters 1996; Abe et al. 2002). In turn, these studies are complemented by research on modern butchery (Ashbrook 1955; Gerrard 1977; Mettler 1987), the role of the butcher through time (Chaudieu 1966), the history of the meat trade (Jones 1976; Rixson 2000), and the development of the modern slaughtering system and industry (Lee 2008; Pacyga 2015), as well as ethnographies focused on faunal exploitation (Cartmill 1993; Howell 1996; Valeria 2000).

The brief overview above makes clear that cut marks have been central to a large body of research, across methodological foci, and with far-reaching conceptual implications. Building from this, this chapter provides a thematic overview of the development of butchery and of the significance of this body of data for how we understand animal utility within historic archaeological contexts. It delves into the nature of information that this dataset reveals. How have analysts used butchery data?

In general, when it comes to analysing butchery, variation between sites and periods has been a major focal point of research. In particular, analysts have paid particular attention to the ways in which archaeological evidence of military presence, intensified trade, and technical variability can reveal broader cultural conceptual changes regarding the commodification of animals. However, while social dimensions have received attention (see later), certain nuances remain to be explored. For example, adopting Bourdieu's *habitus*, Choyke uses research on bone tools to argue that 'marked and sudden changes in the way people chose to make their tools surely signals the advent of significant social change, whether in the form of new ideas or actual movement of new populations into the area' (2013: 9). This stance signals an important point for butchery studies: we can observe the movement of people and things, but variations in butchery practice also illustrate transitions in concepts and ideas.

The themes discussed in this chapter are broadly aligned with those covered in subsequent chapters: economic, social, and technological features. It is also tailored in specific ways to find common ground between the historic context of cut mark analysis (for example, in terms of how taphonomy has driven cut mark recording in the past) and how recording can be adapted to better register process (developed in Chapter 9).

ACCURATE RECOGNITION OF CUT MARKS

The task of the zooarchaeologist is complicated by having to sift through bones from a variety of activities, which contain a range of marks. This foundational observation points to what is simultaneously the main strength of, and yet most complicating factor in, cut mark analysis: variability (Maltby 1985a, 1989). On the one hand, zooarchaeological analysis of cut marks makes possible a wealth of information about every world region and time period, highlighting religious, commercial, and economic preferences. Inferences on carcass processing, utilisation, and butchering, alongside data that evidence trade and settlement economies, can all be gained from butchery analysis (Maltby 1985a: 19–31). However, this depends on distinguishing cut marks from other taphonomies.

As outlined in Chapters 2 and 3, the role of taphonomy in cut mark analysis has been extensively discussed (Speth 1991; Lyman 1994: 1–11; 2010; Outram 2002). For faunal assemblages, the concept encompasses the processes that impact on animal bone, as it moves from the biosphere to the lithosphere (Efremov 1940: 85; Bartosiewicz 2008a). Taphonomy is concerned with two main processes: those that introduce bias or information loss and those that effectively add to the archaeological record, such as culture (Lyman 2010). Analysts have spent considerable effort distinguishing between pre-depositional processes, such as butchery, and post-depositional, such as trampling. The taphonomic framing is most useful in dealing with those processes that could be either pre- or post-deposition, such as fragmentation, which may occur as part of butchery or from trampling (O'Connor 2008: 20–1, see table 3.1 and fig. 3.1; Atici 2006).

Thus, researchers have long been concerned with questions of accuracy in identifying marks (Lyman 1994: 294–315; 2010; Bartosiewicz 2008a; Haynes & Krasinski 2010). Post-depositional taphonomic damage poses an important practical problem for accurate identification of butchery, with factors such as erosion and animal activity playing a particularly detrimental role (Noe-Nygaard 1977; Hill 1979; Dorrington 1998: 109).

Early work was focused on fragmentation patterns as an indicator of butchering behaviours. However, while it has been documented that specific 'spiral fracture' breaks can be achieved using stone tools (Sadek-Kooros 1972; Stanford et al. 1981), it has also been shown that carnivores and the effects of trampling can produce similar breakage (Binford 1981: 78). Thus, even though fragmentation is crucial for studying the context of deposition within and across archaeological sites, and can therefore be important for looking at social space, it cannot be directly equated with intentional bone breakage (Aird 1985: 5–35). Butchery results in bone fragmentation, but so do many other taphonomic factors (Noe-Nygaard 1977; Olsen & Shipman 1988; Todd & Rapson 1988; Maltby 1989). The utility of fragmentation alone for inferring on cultural practice – e.g., to understand butchery guided by religious ideology – is limited. There are simply too many external factors that affect the outcomes. This influences butchery analysis as it throws into doubt the use of quantitative measures as the sole means of securing data from cut marks.

Despite these caveats, Blumenschine and colleagues, working on tapho-nomies that resulted from fragmentation, demonstrated that it was possible to distinguish between cut marks made with metal tools, percussion marks made using stone tools, and carnivore tooth marks, with as much as 99 per cent accuracy (Blumenschine et al. 1996). While these findings are significant, they also need further contextualisation. These three categories are distinct and easily differentiated. Accurate recognition thus remains a principal concern. Shipman carried out influential research specifically on the recognition of butchery (Shipman 1981, 1986; Shipman & Rose 1983). Her research identi-fied a clear distinction between marks left by stone tools and those by carnivore teeth. However, Shipman has been criticised, as the cut marks she analysed were created deliberately for the purpose of leaving a mark on the bone, not resource extraction or butchery (Haynes 1991: 163). Greenfield's pioneer-ing research (1999, 2000) highlighted the differences between stone and metal tools. Stone tools tend to leave a 'U'-shaped groove, and metal knives leave a relatively clear 'V'-shaped indentation with more uniformity and pronounced striation (Greenfield 2000). Thus, identifying metal-tool butchery is a rela-tively straightforward process.

While analysts now have good protocols for distinguishing between cut marks and other taphonomic signatures (Greenfield 1999, 2006; Bello 2011; Schmidt et al. 2012; González et al. 2015), we have yet to deal with a still more basic issue: how to separate what is, and is not, useful for further interpretation (see Chapter 9, and extended case study). Experimental replications can be used in tandem with taphonomic assessment to resolve some of the issues that fragmentation poses, and indeed, there is a well-established precedent in the use of replication studies to improve interpretation of butchery patterns, as demonstrated by Jones (1980) and van Wijngaarden-Bakker (1990: 167–74). However, while Jones appropriately employed stone tools to replicate marks made by prehistoric humans, van Wijngaarden-Bakker used modern tools to replicate medieval butchery. The fundamental relationship between imple-ments and cut marks cannot be overlooked (cf. Grant 1987) and is an import-ant consideration for replication studies. A potential path to better exploit the butchery record, I argue, requires a more comprehensive reassessment of our approach to cut mark record. Shifting our expectation away from cataloguing marks and designing our recording system to instead assess 'process' offers analysts an opportunity to distinguish useful from superfluous marks during the process of data collection.

The research described above primarily focused on distinguishing cut marks from non-butchery impressions. There are a number of options for ways to build on this early work in order to accurately identify the underlying activity, particularly for metal-tool butchery. One field of research that would benefit from more attention is the recognition of different knife-mark typologies. Virtually all enquiries on differentiation between tools have centred on lithics, or between lithics and metal. Little attention has been devoted to the variability between different types of metal tool marks specifically. This is critical for a clearer assessment of process.

Another foundational principle of current butchery studies, with an established precedent, is the linking of location and frequency in order to understand function and process (Hambleton 2013). This approach has been used to study early human butchering (Merritt 2016) and to highlight activities such as skinning, meat removal, and gross disarticulation (Maltby 1985a). Guilday et al. (1962: 63) noted that identification of butchery must take into account repetition of marks at specific locations and that cuts should relate to a function or purpose.

Unfortunately, we do not have adequate qualification of 'repetitiveness' in butchery analysis. It is therefore problematic to suggest repetition can be used for identification without a standard, based on ethnographic case studies, for example, against which location-specific recurrence can be compared and interpreted. Guilday et al. indicate that repetition for the sake of repetition is enough; in fact, a too-simplistic view of repetition can obscure interpretation. For example, can repetitive marks found at certain areas of a carcass from one period be transferred to another period and still provide the same information about butchery technique? This may well be the case; however, at present we do not have sufficient methodological standardisation to be able to make this assumption. Furthermore, our current interpretations of cut mark clusters tend to fall under three main categories: skinning, disarticulation, and filleting. This is a good starting point but it is possible to analyse functionality on the basis of speed, carcass orientation, regional preferences, religious/ritual requirements, and aesthetics, as areas of further investigation.

Accuracy in differentiating cut marks from other taphonomies has been a major driver in past research and will remain critical to the future of butchering studies. However, as the example from Toruń Town illustrates (Fig. 7.1), the potential for more refined and nuanced assessments from the butchery record will depend on individuation between metal-tool categories. In this way, the identification of the cut mark and the analysis of butchery form part of a hierarchical process. The example from Torún demonstrates the power of butchery to serve as an anchor, around which additional findings, from

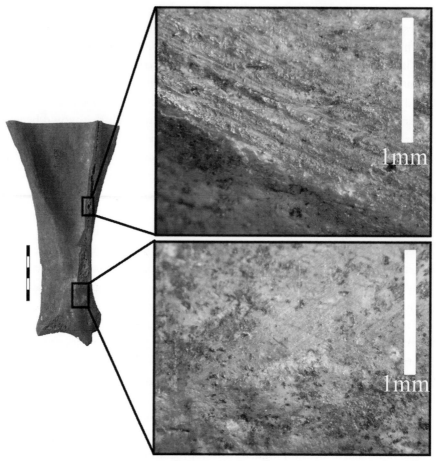

FIG. 7.1 Scapula with two distinct cut marks from the large medieval trade centre of Toruń Town, Poland (German Thorn). The nuances of the striation patterns on this cattle scapula indicated two stages of processing: carcass butchering (lower inset) and preparation for bone working (upper inset). The bone was probably a discarded blank.

metallurgy for example, can be knitted together to reveal details of an animal's death history from its body parts: the use of its bone for tools; the specific parts discarded, refined, and converted into a new object with an entirely new biography. This example also emphasises the need to enhance the taphonomic approach to include the developing principles of *chaîne opératoire* to better assess bones with metal-tool butchery, especially for later periods, which are often categorised by evident differentiation of activity on a large scale.

WHAT CAN WE LEARN FROM THE BUTCHERY RECORD?

Bones from archaeological sites represent the waste from a number of activities. They are the outward manifestation of exploitation for secondary products, consumption, and the discard of bone industries that in the past created many of the products that today are made out of plastic (Boon 1957: 195; Greep 1987: 3). Therefore, artefacts from industrial activities such as horn working, hide preparation, and glue manufacture may potentially be recovered along-side the rejects from household and trade butchery waste (Maltby 1985a: 26; Dobney et al. 1996: 23; Gidney 2000: 170; Noddle 2000: 238). If the distinction between these activities can be accurately deciphered, it would be possible to assess individual activities more clearly (Maltby 1989: 91). Cut marks from archaeological contexts provide evidence of economic, social, and techno-logical contexts. These broad topic areas are discussed in general terms in the remainder of this chapter, with an added goal of situating the causes of variation.

Economics and Butchery

Whether viewed from the methods zooarchaeologists engage with, or the types of interpretation they produce, the economics of animal exploitation has been the major focal point (Grant 2002). This is particularly true for zooarch-aeologists working on later prehistoric and historic sites, where the sites themselves may well have been economic hubs (such as Coppergate, the Viking settlement in York, O'Connor 1989b). Why the emphasis should have been on economic and not social drivers is largely due to the types of questions that early analysts were interested in answering, driven by a parent processual discipline that was itself focused on understanding the economics of past cultures (Clark 1952; Higgs 1972).

For historical zooarchaeologists, the types of large assemblages with which they worked and the types of marks they encountered also played a role in their focus on economic questions. Early studies emphasised the degree of contrast between the cut marks noted from earlier periods, the Iron Age to Roman transition, for example, viewing this as an essential departure from a more deliberate and measured approach to butchering (Grant 1987). A less obvious reason is that the assemblages are in some ways recognisable. They mimicked what one might anticipate could have been produced a few decades or a century or two ago. Until mechanisation, relatively little change took place in the physical processing of animals following the initial use of metal tools (Rixson 2000: 45–60). Thus, the butchery evidence recovered from urban Roman and medieval sites bears a resemblance to modern butchery in terms of similarities between tool marks. One could imagine cleavers and knives used for carcass processing. Indeed, for Britain, the profession witnessed its natal beginnings during this latter period, which continues to influence the modern profession (Jones 1976). Furthermore, the study of cut marks from historic zooarchaeological assemblages has benefitted directly from modern analogy for some time. Rixson, a professional butcher with many years of experience, provided one of the early systems of classification of butchery. Analysts adopted his work (Jones et al. 1985: 136), and he is widely cited both directly and indirectly, whenever reference is made to the three stages of butchery: primary, secondary, and tertiary (Rixson 1988).

Under the overarching umbrella of economics, inference drawn from the butchery record extends from the localised to the regional. In the former case, for example, at both Exeter and Norfolk, differences in butchery practice were suggested to depend on whether the carcass was intended for immediate consumption or transportation (Maltby 1979: 87; Jones et al. 1985: 151). At a regional level, one of the more important conclusions drawn for Britain, and the continent, is the development of a distinct trade in meat based on the systematic and repetitive nature of the cut marks (Maltby 1979, 1985a; King 1991; Hesse 2011). To this, we can add that the techniques themselves may well be 'trade orientated' and developed as a result of increases in demand, driven by an intensification in meat exploitation. From this vantage point, it has been proposed that butchery data may have utility for improving our understanding of complex military, political, economic, and social networks (Maltby 1989: 91; 2007, 2010, 2015; Stallibrass & Thomas 2008). Ultimately, we are able to observe the ways in which demand and trade serve as catalysts for specialisation. This has led to the type of regional and diachronic comparative assessments that are a particular strength of zooarchaeology. A number of analysts have synthesised data from bone assemblages across periods to highlight the main differences in animal exploitation, from the Neolithic to the Iron Age (Wilson 1978) and, in particular, between the Iron Age and Romano-British period (Maltby 1981, 1985a; Grant 1987).

We can turn from this discussion, focused on how the butchery record can reveal practices of animal exploitation in the past, to consider how future work might utilise cut marks to understand local and regional economy and transitions over time. Even in its most basic form, processing meat is an activity dependent on implements. Consequently, a distinct meat trade, as with any other, will require its practitioners to cooperate with a wide range of other professionals. For example, interactions with blacksmiths would need to take place for the production of cutting tools, and carpenters might be required to produce butchers' blocks, display counters, and hanging equipment for salting or dry curing. Other relationships could develop with specialists in allied professions that depended on the butcher to deliver their raw material, such as bone workers (Serjeantson 1989). Thus, we observe aspects of cross-trade socioeconomic links and how the changing fortunes of one profession impact the activity of another (Yeomans 2007).

It is also worth exploring some of the less obvious aspects of trade and demand. Establishing whether aesthetic considerations were taken into account, either through cut marks or from iconographic and written sources, could lead to a better appreciation for how the meat itself was perceived. Aesthetics serves to move our assessment away from a model of carcass processing based purely on nutrition to one that incorporates commercialism, status, and monetary value. Effectively, this attempts to establish the 'retail preference' for the site in question. Using a modern analogue, within the last twenty years 'prime cuts' for the modern Western palate invariably refers to meat with a relatively low fat content or with little if any bone, indicating specific cultural biases that are not founded solely on nutrition. By assessing butchery in its wider economic contexts and broadening the interpretative framework, it is possible to reconsider the role that butchery has played in the development and perpetuation of economic innovation and intensification.

Butchery as a Lens on Culture

The skeletal morphology of an animal exerts a powerful influence on the pattern of butchery (for example, size), but there is still a great deal of flexibility in how a carcass can be dismembered. This flexibility is often articulated as regional, religious, or ritual variation and results in discrete differences, expressed on the bones themselves (Moore et al. 1983: 11–38; Maltby 1989: 75–107; King 1999; Swatland 2000: 56–63). Ethnographic and historic research emphasizes the cultural role of meat preparation and consumption (Fiddes 1991; Mooketsi 2001; Grant 2002). The anthropological literature offers numerous case studies where meat processing has been used to deepen our knowledge and understanding of a cultural group (i.e. White 1952, 1953, 1956; Gould 1967; Yellen 1977: 271–332; Binford 1978, 1981; Brain 1981; Bunn 1981; Speth 1983; Bunn & Kroll 1986; Bunn et al. 1988; Miller 1995). If we approach the subject from the perspective of the wider role of meat in society (Fiddes 1991), butchering itself can improve our understanding of food sharing (Mowat 1970; Mooketsi 2001), cuisine (Goody 1982), gender roles (Gifford-Gonzalez 1993), and divisions of labour (Ferguson & Zukin 1995). While these topics have a more direct bearing on consumption within social settings, butchery has played an important role in the expression of ritual, ideological, and religious identity (Chakravarti 1979; Wilson 1991: 92–3; Brumberg Kraus 1999).

Binford conveyed the idea of cut marks serving as a lens on culture most effectively in his description of butchery as the 'dynamic expression' of 'cultural rules' (Binford 1978: 47). Accurately interpreting the dynamic expression – which here I have taken to imply the techniques of processing – depends largely on our methods for recording cut marks. The cultural rules could serve as a guide to our interpretation, and it is these elements that I emphasize here. For archaeologists, many studies have drawn on the rich bone assemblages from late prehistoric and historic sites to assess human–animal relationships (Billington 1990; Hassig 1995; Thomas 1999, 2006; Pluskowski 2007; Sykes 2007). However, with exceptions (e.g., Grant 1987), butchering has not usually played a role in how we interpret non-subsistence aspects of human–animal interactions. This is unfortunate, as butchery datasets are effectively repositories of social attitude, from a number of different perspectives.

The most obvious example emerges within the context of ritual and reli-gious expression, specifically, with identifying sacrificial slaughter and the subsequent use of sacrificed animals (Cope 2004: 25–7). Achieving a clearer understanding of ritual activity is dependent on the recognition of specific cuts, distinct from those that would be produced during carcass processing for non-ritual subsistence. Herein lies the problem: this may not be as clearly defined as one might assume. There is bound to be a degree of overlap between quotid-ian practices, ritual slaughter for feasting, and sacrifice. For archaeological contexts, other concerns need to be taken into account. For example, some scholars have suggested that consumption of meat occurred only after ritual sacrifice in the Roman world (Belcastro et al. 2007). Although this statement needs to be contextualised within both regional and chronological settings, given the complexity of what constitutes 'Roman', it does give rise to at least one important question: can a better understanding of meat in the diet also provide a clearer assessment of ritual practice in everyday life?

Ritual and religious practice is closely identified with the initial cut used to extinguish life. For followers of Islam, for example, a number of additional considerations have to be abided by. Although not universally mandated, some Muslim groups insist on the animal being fully conscious. Furthermore, Sharia law extends well beyond the act of slaughter; the animal should be raised such that it is allowed to behave in a normal manner. The person slaughtering the animal also comes under scrutiny. Both men and women can slaughter, but they must be of sound mind, preferably be Muslim (though Muslims can eat meat slaughtered by Jews or Christians), and have attained the age of discretion. During the act of slaughter itself, the animal must face qibla (the direction toward the Kaaba in Mecca) and be slaughtered with one stroke of the knife. The blade to be used for slaughter cannot be sharpened within sight of the animal, nor can one animal be slaughtered within sight of another (Fuseini et al. 2016).

A defined example of religious practice within an archaeological context can be found in the case of kosher slaughter (Valenzuela-Lamas et al. 2014). More than one cut forms part of the suite of processes that must be undertaken for the animal to be considered kosher (Greenfield & Bouchnik 2011: 106). Cope's research describes this form of slaughter in terms of the practical tasks involved. Her work draws on examples that illustrate the initial cut for slaughter, as well as subsequent marks that appear in locations that both correlate with Judaic text for proper butchery and would be unlikely to occur as part of everyday, non-religious, practice (Cope 2004: 29–31; Bouchnik 2016: 306). Cope's example also shows that detailed study of the cut marks reveals important nuance: the slaughtering stroke was undertaken once, with a sharp, well-honed implement (described in Chapter 6). Furthermore, richness is added when closer attention is paid to the butchery process: marks show the sciatic nerve was removed, as evidenced by a cut to the internal (medial) facet of the thigh-bone (femur) (Cope 2004: 30).

Other examples point to the benefits of connecting the cut marks to the larger animal bone dataset. By knitting the precise evidence from butchery marks into the global faunal assemblage from a site, it is possible to identify preferences for animals used in ritual practice based on age, sex, or other defining morphological features (e.g., the shape or size of horns) (Chaix et al. 2012). Taken as a whole, the butchery record provides strong evidence for religious practice, and while it may be necessary to corroborate the details of the cut marks with written sources or iconography, clearly the potential exists for direct identification of ritual and religious slaughter from the marks themselves.

Another point to consider is that social and economic factors need not be studied as discrete topics. Intensification in trade leads us to consider how economic drivers influence the place of animals within a given society. Is it possible to draw conclusions regarding how domesticates were viewed based on changes in demand for meat or hides? We might attempt to decipher the chronological moment at which the perception of animals begins to take on new meaning: the commercialisation of animal bodies (Seetah 2005: 6; Sykes 2014: 15). A working hypothesis might suggest that the intensification in demand – for animal traction or meat – had a profound impact on the way animals were treated and consequently viewed.

For example, indications of improvements in cattle morphology (Dobney et al. 1996; Murphy et al. 2000) and the increasing importance of traction and dairying (Maltby 1981) in Roman Britain points to the fact that cattle had a different role in the day-to-day lives of people during this period, when compared with the Iron Age. The same activities took place: the exploitation of secondary products. However, the details are fundamentally different, feeding the pursuit for nuance that drives archaeological research. This type of interpretative assessment, using economic details to guide our understanding of social attitude, is ideally carried out in tandem with wider assessments at the settlement level. Information connecting husbandry, building structure, and the way animals were processed needs to be brought into a cohesive framework that illustrates how changes took place in intensity of animal management, architectural design to house the animals, and techniques of butchery. This then provides the structural framework for a more informed assessment of the place and view of animals in past cultures.

Technology of Butchery

The intrinsic links between butchery, butchering, and implements is an important one. In terms of quality of material, this is one area where the archaeological record could potentially be as rich as the ethnographic. However, there is a limitation to consider. The ethnographer has the ability to view the application of the tools in question, witnessing the techniques used for butchery first hand. Clearly, this is not possible for the archaeologist, although usable correlates can be developed through experimental replication.

The archaeological study of butchery tools and techniques falls into two distinct subdisciplines, archaeometallurgy and osteoarchaeology.[1] In practical terms, these areas of expertise deal with their material resource in different ways, and increasingly, metal and bone are studied at ever-finer levels of resolution. For metal finds, this involves detailed studies characterising the chemical, thermal, and mechanical properties of the object, based on light optical and electron metallography of the microstructure (Wayman 2000). Bone, on the other hand, is the subject of complex molecular assays. At an even more basic level, the artefacts and ecofacts themselves are unlikely to be excavated in close association, and once recovered and analysed, will not meet until they appear as separate sections in published works or site reports. This is the archaeological process, and I am not suggesting we change it, nor am I advocating that we necessarily develop expertise in each other's field. For the metallurgist, understanding how bones were cut has limited relevance. The onus, however, does fall on the zooarchaeologists, in particular those concerned with gaining a clearer understanding of butchery. As Greenfield (2013b) has shown, faunal specialists have much to gain by integrating across these fields. However, while the benefits may be self-evident i.e., the ability to visualise the interplay between these two branches of daily life, the reality of integrating data from recovered metal and bone finds is not straightforward.

In an ideal situation, the analyst would be able to correlate cutting tools from a given site with different mark typologies, following the principles of forensic analysis. The striation pattern of a mark can be as distinctive as a fingerprint, and indeed, it is this principle that underpins criminal investigations where a knife has been used as a weapon (Lewis 2008). However, with the exception of raw bone destined for tool production, itself a universally important craft and industry (see Sidéra 2013; Campana & Crabtree 2014; and contributions in Choyke & O'Connor 2013), animal bone is invariably a waste product, while metal-tools are valued objects, retained and transported from site to site. Thus, the integration by necessity has to take place at the regional scale. General reference to tool types can be connected to patterns of butchery. An example would be to look at the relationship between increasing demand for meat and changing trends in the tools used for butchery. This would be readily evident from the butchery record and would mimic the type of study evaluating the adoption of metallurgy (Greenfield 2013a).

Moving from tools to techniques of butchery, the details of practice can improve our understanding of how implements were used, helping to discriminate between types of activity. As with the tools, demand contributes to variation in the butchery record. Increased meat consumption, in terms of quantity (and, if it can be deciphered, quality) can serve as a catalyst for the development of new modes of butchery. Furthermore, demand serves as a catalyst for technical and implement specialisation (Seetah 2004, 2005: 4–5). By tracing the diffusion of the technical features of butchery, we may be able to investigate the processes of knowledge dissemination from region to region, of a strictly technical nature (Marciniak & Greenfield 2013: 441–2; Seetah et al. 2014: 65–70). The preceding places emphasis on the methods and tools of processing. However, these do not represent the totality of technical aspects that form part of carcass processing. Additional techniques relate to how the carcass was hung, stored, and exsanguinated and the methods used to mitigate spoilage or prevent *livor mortis* (also known as hypostasis, referring to the pooling of blood in the dependent part of the carcass after death). None of these aforementioned actions actually requires a single cut mark, but forms important aspects of butchery.

In some cases, these technical and logistical considerations have further ramifications. Using storage and building structure as an example, by identifying techniques that are analogous with trade and by noting structures that indicate reinforcement of the building (or areas of the building) that allow for a carcass to be hung, we have the grounding to establish whether premises were used for butchery and storage of meat (Seetah 2004). This type of inference remains elusive in archaeological research and would add considerably to our ability to place the craft within a wider framework describing its mode of operation. Certainly for the Romano-British period meat markets (*macelli*) and abattoirs (*lamenae*) were known to exist and have been noted from the archaeological record (Gerrard 1955: 2; Maltby 1989: 75). Thus, the technical aspects of butchery and the principles employed by the butcher are established from the bones. The osteological findings are corroborated and bolstered through evidence supplied from other branches of archaeology, as well as iconographic and written sources. We are able to construct a better understanding of how the craft functioned, adding dimensions to a series of practices that could conceivably be considered self-contained in terms of their implications for other areas of trade or society.

CONCLUSIONS

In discussing economic, social, and technological features of butchery, the underlying message has been that, when focused on historic societies, analysts often deal with variation on a scale that can complicate accurate interpretation. Although this variability leads to methodological concerns in terms of extracting evidence, it also provides scope and opportunities for interpretation. This is well supported by the impressive body of research that has placed butchery within context, with clear appraisals that underline the agency of consumption.

In turn, this has led to important insight into both economic and cultural transitions across broad time spans. Maltby (1984) situated the role the army played in the development of a 'meat market system' in Roman Britain, while Dobney et al. (1996), reporting from Lincoln, East Midlands, studied specific butchery techniques from the scapula to suggest transitions in curing techniques. Grant (1987) analysed variations in patterns of butchery from the Iron Age to medieval period in Britain, identifying and linking specific tools with particular butchery trends. Grant demonstrated that both Iron Age and Romano-British butchery showed distinct 'period specific' uniformity, each with its own tool-kit of butchery implements. These innovative studies highlight the possibilities for interpretation.

This chapter underscores the need for a multidisciplinary approach to cut marks (Seetah 2006: 112). It is through this approach that we are able to place butchery data into a more holistic and accurate framework. As this chapter and other authors have argued (Grant 1987), butchery can provide keen insight into economy and culture. However, the social inferences lag behind, and despite the fact that ethnographies illustrate the significance of butchery beyond economics, relatively little effort has been made to make this association more concrete. This unequivocal indicator of human activity should lead to more than a basic understanding of aspects of meat processing (Grant 1975: 383; Noe-Nygaard 1977; Aird 1985: 5).

In conclusion, I want to underline two larger issues: the first concerns the question of primary data recovery, namely, how can zooarchaeologists deal with the recovery bias and its influence on the butchery record? Looking to the future of cut mark analysis, any general overhaul of how we study butchery will need to closely examine the parameters that directly and indirectly influence variability in the butchery record. As others have pointed out, factors such as the skill of the butcher or availability of osteological expertise during excavation influence our ability to assess the butchery record. The second issue is more problematic: the process of data recovery may fail to record or report on butchery at all, even in cases were animal bones are analysed. MacKinnon's research (2004: 163), drawn from an extensive survey of Roman sites in the

Mediterranean, indicated that some two-thirds of the bone reports included in his study failed to provide any evidence of cut marks. The next chapter describes how our interpretation of butchery is negatively impacted by current methods. However, MacKinnon's valuable work places in stark relief the extent to which the lack of a standardised approach to cut mark analysis continues to hamper our ability to draw a more complete and nuanced picture of human–animal interactions in the past.

CHAPTER EIGHT

PROBLEMATISING BUTCHERY STUDIES

INTRODUCTION

Based on the review provided in the preceding chapter, it should be clear that the historical and anthropological study of butchery has largely been synonymous with cut marks. It is 'the cut mark', defined taphonomically, that has provided the conceptual bases for our analytical approach. But what have been the implications of emphasising the cut mark? If a lack of standardisation represents a structural problem, other issues relate more specifically to how we record cut marks: what are we actually representing with the outcomes of our current methods?

Furthermore, a schism exists between the recorded dataset and the component of that data that is conveyed in site reports or published works. Despite the fact that we record many hundreds – if not thousands – of cut marks from an assemblage, only a limited cohort is used to infer a pattern of butchery. Often, these few marks are then represented in a schematic format. Are we making the best use of the global dataset?

The degree of differentiation between categories of butchery, e.g., kitchen versus trade, highlights the variability inherent in the bone assemblage. This can be a boon for interpretation, as it offers many avenues for thinking about diversification and specialisation of crafts and about location-specific patterning. However, variability can also be problematic in that it can be difficult to differentiate between the various influences, taphonomic and anthropogenic (though these are not mutually exclusive), which have helped create the assemblage. Current methods tend to exacerbate the problems of variation, as each mark is ascribed equal weighting. There is also a tendency to represent all the cut marks from an assemblage on a single diagram, increasing the likelihood that the marks are conceptualised as a singular entity. As a result, the analyst is effectively concatenating a range of activities that represent butchery occurring in diverse locations, for a variety of purposes, by different people, with a range of skill levels. The compression of archaeological materials from across significant time ranges further compounds this already complicated situation.

Constraining what we record to specific aspects of the dataset – e.g., to geometry or type of mark – inadvertently, and I would suggest negatively, influences interpretation, particularly with regard to the tools and craftspeople. Evidence for this is presented below, based on a survey of zooarchaeological literature. The principles of *chaîne opératoire* described previously encourage analysts to situate the practice, craft, and craftsperson within their broader contexts. This is important not only for zooarchaeology but for archaeology overall, as it reminds materials analysts to conceive and examine the objects we study within larger societal frameworks that would have surrounded the pot, knife, or vase during the lived experience.

This chapter first identifies a suite of methodological limitations that shape the challenges of studying butchery as a craft-based practice. Following this, I examine the implications for interpretation: how do current approaches to cut mark recording hamper the ability of analysts to describe past human–animal interactions, production processing, and consumption practices? It would be unreasonable to expect universal consensus in the way analysts have inferred from this dataset. However, identifying shortcomings in previous research provides a basis for reshaping both the analytical and inferential frameworks used to study butchery. Why have gaps occurred? How might we resolve some of these shortfalls?

INHERITED LIMITATIONS IN THE ANALYSIS OF CUT MARKS

Despite numerous attempts to establish a suite of protocols that all zooarchaeologists could follow, global adoption of standard methods has not materialised (Baker & Worley 2014). The benefits of standardisation can be seen with the widespread application of measurement protocols, which greatly facilitate comparisons between sites (Von den Driesch 1976). In the same way, standardising the recording of butchery facilitates comparisons, not only between periods and at a regional level but also between different tool technologies, such as lithic, bamboo, shell, or metal. Lauwerier's important research on Dutch Roman sites aimed at providing a standardised protocol to record butchery data. His work set out a number of guiding principles for recording cut marks and was supported with a large appendix of schematic examples (Lauwerier 1988: 40–2, see also appendix, 181–213). However, while theoretically informed and methodologically well structured, in the absence of experimentation to support the model, the protocol was detached from the activity inherent in the craft. Standardising how we record butchery need not be onerous. As a starting point, analysts already use a language to describe the recorded data: chop, slice, and knick, for example. Constraining these terms in specific ways could provide the basis for standardisation of terminology and for describing behaviour. Similarly, the developments advocated in Chapter 9 to facilitate the recording of cut marks are based on traditional concepts already established in zooarchaeology. These include computer-based data capture and analysis, using forms in software such as ACCESS or FileMaker. Approached in this way, the call for standardisation is not a recapitulation of a more general appeal addressed to zooarchaeology (Baker & Worley 2014). Rather, this is an invitation to make slight modifications to our current systems and terms, leading to a sustained adoption of the new approach, with attendant improvements in interpretation (Maltby 2014: 35–6; Seetah in prep a).

One of the limitations of emphasising cut marks is that we have tended to study them from a taphonomic standpoint, as discussed in Chapter 3. Analysts have focused on whether we are accurate in distinguishing between cut mark typologies or between cut marks and other taphonomies (summarised in Lyman 1994: 297). If this represents the conceptual basis from which our methods flow, the main limitation with our current recording protocols is a reliance on alphanumeric coding. These systems, such as BoneCode (Klein & Cruz-Uribe 1984: 115), can be traced to the advent of computers and were initially developed to rapidly record bone assemblages (Gifford & Crader 1977).

These same systems, which were not immediately intuitive, were adapted for use in cut mark analysis. Invariably, the protocols became increasingly cumbersome with the addition of extra parameters to accommodate new cut mark types or locations. There is a fundamental problem with the use of a coding system. In effect, the analyst is only indirectly responsible for creating the outcome. By constructing evidence from the butchery record in this way, the analyst opens the door to the code producing a misreading of the data. Moreover, by its nature, a code homogenises data to fit into specific categories. Thus, the analyst loses complexity.

These are not problems that can be easily mitigated. Computers have been an immense boon, facilitating data capture, analysis, and reporting; analysts will always have to generalise from the data. The concern is that researchers should be making the decisions as to how best to present their data, rather than having this process imposed upon them by the recording system. Analysts commonly adapt observations to find the nearest 'fit' to the established protocols. This can result in substantial additional commentary to the butchery records in order to provide an explanation for discrepancies and deviations from the norm (Maltby pers. comm.).

Current recording systems have a tendency to restrict analysts, constraining what is registered to specific parameters, such as location of the mark or its type. Effectively, they were *designed to focus on the minutiae*. They do not encourage interpretation, nor do they lend themselves to investigating the underlying butchery process. These factors further complicate an already complex data resource and do little to assist in capturing and describing the important aspects of cut marks.

Other issues have served to compound the problem of design in our recording systems. The approach to cut mark analysis has been developed from assemblages that were created using lithic implements or modern-day ethnographies of aboriginal groups, who themselves often used lithics (e.g., Guilday et al. 1962). These same methods developed from lithic-tool butchery were adapted for use on later prehistoric and historic sites, where metal tools were employed.

Using bone from assemblages created with lithics is constraining as these tend to be less well preserved and smaller. More importantly, assemblages created with lithic as opposed to metal tools differ from each other in significant ways. At the site level, the formation processes that create the assemblages, and conditions within stratigraphic units that contain bone, are themselves different. The proportion of bones with cut marks is usually higher where metal tools have been used. In part this is because marks from metal knives are more distinctive (Greenfield 2016: 278).

Furthermore, by the time we see the widespread adoption and dependence on metal, we are dealing with later periods, often with better-preserved assemblages. These same assemblages can be very large, in some instances numbering many tens of thousands of bones and occasionally derived from waterlogged conditions that are ideal for preservation of bone. The socioeconomic influence, as well as the scale of commercial activity, is reflected in the bone assemblages (as noted from Viking York, O'Connor 1991, or medieval Novgorod, Maltby 2013).

Across these cases, researchers tend to be interested in fundamentally different types of questions. Investigations of cut marks with lithics, from the 1980s onward, have been concerned with demonstrating human agency and the differences in tool use between ancestral hominins and anatomically modern humans, or between humans and other tool-using animals (Ambrose 2001; Luncz et al. 2015). In essence, these studies were driven by research focused on human origins and evolution. While the study of cut marks is also deeply relevant to this important subject matter, the interpretative framework and questions that this research generates are quite different to the types of enquiry that form the focus of research on later, historic periods.

FIG. 8.1 Cut mark made by a fine-bladed implement on a pig axis. Note the raised but intact edge of the bone (Riekstu hill fort in Cēsis, Latvia). This type of mark would be impossible to make with lithic tools, even those with vitreous properties that can retain a keen edge. Obsidian, for example, would be too brittle to penetrate the bone to this extent without breaking the cutting edge.

The difference between how these two tool categories are created and used also requires consideration. Cutting tools are a critical component of how the knowledge of butchery is enacted. Implements are also an important factor in terms of creating variability. The significance of tools thus goes beyond the recognition of tool type that inflicted the marks (Greenfield 1999, 2000); scrutiny of tool use provides a range of insights, from the nuances of the techniques used in butchery to the adoption of new technology (Greenfield 2013b). The more important point examined here is that metal tools make fundamentally different types of cut marks than those that can be created with lithics (Fig. 8.1).

Metal is more malleable than stone (see discussion on fabrication in Chapter 6). This allows for greater control over the functional properties of the final product as well as greater variability in the types of implements that can be created. Metal tools can be more easily refined for specific uses. Functionally speaking, there is a smaller suite of cutting techniques that can be undertaken with lithics (Greenfield 1999). Furthermore, when using lithics, the animal's skeletal morphology places stricter limits on how the butchery can be executed (Walker 1978; Seetah 2008). With the creation of large metal blades, and in particular cleavers, this constraint is removed. Indeed, this may have been the catalyst for the creation of specialist tools. Thus, the butchery process is altered substantively, and, it can be argued, this was undertaken by design – and specifically, the design of the tools.

Taking these points into account, an updated recording system has to accommodate the greater variety of cut marks that can be made with metal tools, providing scope for a broader inferential framework. In essence, the current approach is counterintuitive. The methods in use today were created from bone with a limited suite of cut mark types, often from poorly preserved and small assemblages. It would be intuitive and productive to employ the much larger, better-preserved assemblages, with easily identifiable cut marks that show greater variability, to construct our recording systems.[1] These points also help explain why much of the focus of our current research on cut marks has been centred on distinguishing between cut marks and other taphonomies, or clarifying whether the cuts were made with lithics or metal (Greenfield 1999, 2000). These are important concerns for those analysts working on prehistoric assemblages, given the taphonomic context. They are less significant for analysts working on later materials, where it is generally easier to identify the marks.

Critically examining our current approach to cut mark analysis and identifying the problems that emerge under scrutiny leads to a set of important related questions: how can we mitigate the inherited limitations? What steps have been taken in answering the call to move beyond a focus on the cut mark? Why have our methods stagnated? Answering these questions – particularly the last – rests on how we think about butchery. To return to a point raised in Chapter 3, experimental archaeology holds much promise for widening our view of butchery. However, we need to address how actualistic studies are performed in this discipline.

The crucial component of any experimental study is the practitioner. Studies have shown the potential for introducing bias when the practitioner and researcher are the same individual (e.g., see Domínguez-Rodrigo 1997). Alternative options utilise 'naive butchers', i.e., those without butchery expertise, or, at the other end of the spectrum, engage professional butchers. Use of the former is seen as a way of removing bias, as the practitioner cannot introduce any preconceived ideas of how the butchery should be undertaken. In specific cases this is well justified (Domínguez-Rodrigo et al. 2012). However, using naive butchers as a control measure means that primacy is given to individuals with no practical knowledge of the subject, which directly contradicts the fact that in the archaeological record we are observing evidence of people who knew how to butcher. Preconceived ideas – i.e., planned approaches in the mind of the practitioner – drive butchering.

What, therefore, is actually being produced? There needs to be an acknowledgement that the outcome is effectively an artefact and bears no reflection to skilled production, a point long since recognised (Haynes 1991: 163). The mark being created may replicate exactly what we observe, but does it accurately represent the underlying drivers, such as efficiency, or the specific tool and how and why it was used, the dexterity of the practitioner, or which hand they used? More importantly, do the results from these studies help us better understand the decisions that underpinned the actions we observe, knowing that there has been no attempt to replicate the expertise inherent in the butchery process?

Similarly, using professional butchers has potential drawbacks. Asking the butcher to reproduce a mark can introduce bias and inaccuracy; instead, it is the underlying process that needs to be reproduced in order to clarify the steps and wider context of how the mark was made (as in Machin et al. 2005). This relatively small modification may help to mitigate issues of equifinality. Understandably, most analysts will not share the deep knowledge of carcass processing held by a professional butcher. While we might therefore assume that there are numerous ways of producing a mark, there may only be one or a small suite of possible options. This knowledge rests with those skilled in butchery, and analysts would benefit by tapping into this resource whenever possible, making the most of modern analogy (Seetah 2008). However, we also need to exercise caution. Professional butchers may not have experience in using exactly the same equipment as recovered from the archaeological record. In the Western world at least, consumer and health ideologies, as well as safety/ethical demands, heavily influence butchers.

In addition to being conscious of introducing bias into our experimental design, we also need to be mindful of how the replications themselves will clarify the process of butchery. How do the marks from a given assemblage fit within a *schematic* of butchering? Actualistic studies are generally carried out independently and rarely with an explicit attempt to build on previous work or develop a framework within which individual projects can contribute. In contrast, experimental research undertaken to distinguish and clarify cut mark typologies shows a progression from an initial assessment of tool signatures (Greenfield 1999) that become more refined and distinctive (Greenfield 2006), as well as more scientific (Schmidt et al. 2012). This approach has considerable utility for the experimental replication of cut marks and promotes a holistic and integrative structural framework that can accommodate regional and chronological variations.

More recently, in a response to the need for better recording and standardisation, new methods have been developed based on template models onto which cut marks can be ascribed (Abe et al. 2002; Popkin 2005; Orton 2010; see Fig. 2.1, Chapter 2). However, these methods continue to conflate *recording* and *representation* of cut marks. Zooarchaeologists certainly need effective ways of visualising their findings, and these types of systems offer rapid and useful ways of doing this using GIS or other visual recording systems. They facilitate a form of data capture but are not in lieu of accurate interpretation of cut marks. These methods still focus on the location (mainly) and basic morphology of the mark; they currently have no way to record process.

Furthermore, they only indirectly (and I would argue inaccurately) contribute to standardisation, one of the stated goals of these new approaches (Sykes 2014: 15). Template methods may well be standardising the recording of cut marks, but they are standardising a flawed system. Representing all the cut marks from an assemblage on one schematic diagram of an entire animal, or a body part, has considerable utility for comparative purposes, to identify patterns, or types of marks that are repeated with consistency. However, even in these situations analysts are creating an artefact of the data.

IMPLICATIONS FOR INTERPRETATION

Having identified shortcomings in the recording methods, I now examine the ways in which our approach to cut mark analysis has influenced our ability to draw accurate inferences from the butchery record. These start at a general level, for example, an incompatibility of datasets across time periods and regions. This does not mean that variability has not been a topic of study – far from it (e.g., Grant 1987; King 1999; MacKinnon 2004). The issue is that it has not been possible to develop these comparative models to a particularly high level of resolution. Comparisons often derive from single, large, well-stratified sites (Maltby 1979) or sites from one region or type (O'Connor 1989a, 1993). The detailed pan-regional assessments, such as those undertaken by King (1999) and MacKinnon (2004), remain elusive for patterns of butchery.[2] Thus, analysts have drawn relatively superficial conclusions, such as 'traditional butchery of rural regions' (Serjeantson 1989: 5). I stress that the fault does not rest with the analyst. Our methods simply do not allow us to extricate sufficient detail. If we assess butchery only in general terms, we set the bar at interpretations of 'skinning', which is the same task no matter where it is undertaken. Thus, comparison is unnecessary. Discovering the nuances of how and why different tools, techniques, and drivers exist for individual episodes of skinning is a task for future studies of butchering.

Before proceeding, I would like to make it clear that it is not my intention to undermine the work of those authors to whom I refer in the following text. Using published examples serves as a point of departure for discussing gaps in our methods and for drawing attention to shortfalls in our interpretation of butchering. Furthermore, a number of cases of misrepresentation are still evident in recently published research. Thus, we do need to be aware of established problems in our use and interpretation of the butchery record lest they be repeated. The following is not an exhaustive account, but it serves to highlight some of the main areas of concern.

Applying Modern Analogy

Comparisons with examples drawn from the modern world offers much for those interested in a better understanding of human–animal interactions, particularly within the context of consumption. Any benefit has to be tempered by the potential risks of misappropriating analogy. Our own contemporary perceptions of butchery, and indeed, our view of those involved in the profession (see Chapter 4), may indirectly have a negative influence on our interpretation. Given the need for vigilance and building on the background set out in Chapter 7, I would like to outline a number of examples that illustrate potential pitfalls.

Stokes presents an insightful example of the depth of information that can be drawn from the butchery record as it relates to a military enclave (Stokes 2000: 145–52). By correlating the spatial location of where specific carcass units were recovered from the barracks of the second-century site of South Shields, Newcastle, he demonstrated that the commanders were provisioned with 'choice cuts' of meat. A detailed account of the methods used to process the carcass led to a pattern that Stokes refers to as 'block butchery'. Stokes considered this as evidence of 'conspicuous fairness' in terms of portioning (Stokes 2000: 147). This type of overt statement regarding carcass distribution among the soldiers is precisely where the social facets of butchery are revealed, serving as a guide to meaningful inference.

Unfortunately, this important study did not take into account the numerous ways in which implements played a role in how the butchery was performed, a component of the dataset that could have helped to more effectively clarify the techniques of butchery and of the physicality, the embodied actions, of the butchers. Stokes concluded that the butchery was undertaken by individuals with little knowledge of animal anatomy or without incentive to execute the task with due care, further stating that some of the butchery was repetitive and was often very close to joints, as opposed to through the joints (Stokes 2000: 147).

Although not mentioned explicitly (Stokes makes no reference to cutting implements), it is evident that heavy blades were used, as indicated by chop marks. Taking a more holistic view of the dataset, one might suggest that the presence of large blades or cleavers would have significantly influenced the method of butchery (Seetah 2005, in prep c). Cleavers facilitate a type of portioning that does not require the butcher to use techniques to work around bone (Seetah 2006: 112; in prep a), but rather can chop through bone or joints. Describing the potential tools that were used, readily noted from the cut marks, immediately provides a more comprehensive assessment of some of the technological dimensions for the strategy of butchery observed.

Developing this line of reasoning further, the techniques of butchery that depend on heavy blades employ more forceful actions such as chopping. This singularity provides not only a point of comparison, say with small-blade butchery from rural sites, but also helps to better portray the know-how, comprehension of anatomy, and skill of the butcher. When using a heavy blade, a 'prep' mark is invariably required to locate the joint before delivering a chopping blow (see Chapter 9). Chopping into flesh, and not directly onto bone, is ineffective as the soft tissue cushions the blow, dissipating the force of the chop through the muscle, crushing rather than cutting the meat. This helps explain why there were multiple cuts around, rather than through, the joint: the technique involved two steps, 'prep-then-chop'. The butchers at South Shields would certainly have had a good working knowledge of animal anatomy to achieve this action and would have required some skill to perform this form of more forceful butchery. Continuing further in the same vein, this type of butchery invariably requires a butcher's block, as use of heavy blades to chop is best performed against a solid base. This does not reveal details about the quality of the butcher's block, only that such supplemental paraphernalia was likely part of the tool-kit of these military butchers based on the technique and tools used to butcher. The evidence for this more dynamic and vigorous form of processing also lends itself to depictions of the scale of operations; this strategy likely involved the butchery of numerous animals in consignment, perhaps once a week, to provision a meat ration for a period of time.

Finally, a more comprehensive exploitation of the dataset also reveals a nuanced, phenomenological view of the setting and the enacted taskscape: a group of well-trained, skilled butchers, working in a physically demanding environment, processing numerous animals in sequence. Their actions are undertaken to a predetermined plan to produce meat as units that are then fed into the military machine. The meat not only serves as a source of nutrition but also reinforces both social differentiation between commanders and lower ranks as well as equality between personnel of the same rank. It is this constellation of agents, tasks, skill, tools, and location that leads to compartmentalisation of flesh into 'blocks of meat' and defines 'block butchery'.

The South Shields case not only demonstrates a misrepresentation of butchery tools and techniques (see next section) but also reveals that for some authors there is a difficulty in reconciling the aesthetic characteristics of modern butchery with the more functional approach from the Romano-British period (see, for example, Aird 1985, and the section on 'Misrepresentation and Misinterpretation', below). This point reveals an incongruity between the way archaeological butchery is interpreted and the use of analogy from the contemporary context.

In the example above, a modern perspective on animal morphology is projected into the past. The 'block butchery' described does not acknowledge that cattle of the period would have had less differentiation in musculature over the carcass (Dobney et al. 1999). Furthermore, Stokes discusses the likelihood that the commandant was being supplied with cuts such as 'rump, sirloin, and fillet steak' (Stokes 2000: 149). While the commandant may well have received the best portions of meat, there is no evidence to suggest that the butchery process was aimed at producing specific cuts of meat, as in the modern setting. Indications of 'meat cuts' are often interpreted from the faunal record, and in numerous cases these are well supported (Jackson & Scott 2003; Sykes 2007, 2014). Meat cuts have also been used to generalise on patterns of consumption in historic contexts (Lyman 1977).

Here I would like to bring together the conceptual issues raised in Chapter 1 vis-à-vis 'studying bone to observe meat' with the misappropriation of the modern context as it relates to faunal physiology. As the following cases emphasise, much depends on animal morphology and both the theoretical and pragmatic ramifications of skeletal and muscular structure need to be concretely integrated into our analysis of the butchery record. Animals today are raised to have sufficient muscle differentiation to make it feasible to separate different portions of meat, with ostensibly different qualities (Swatland 2000). Thus, when ascribing modern meat joints onto archaeological samples we err, as the animals themselves were not the same. This leads to another area for reflection: animal husbandry. This is a central topic for butchery, as the constitution of the animal is critical to the way it is processed.

For cattle in particular, as one of the most functionally diverse domesticates (O'Connor 2014), we have to assess carefully when improvements commence (Albarella et al. 2008), as well as specialisation of types into breeds (MacGregor 2012: 426) and how these developments relate to primary and secondary product exploitation. This is where historical accounts can be particularly useful (Ritvo 1987: 45–52). Furthermore, husbandry patterns optimised to increase meat yields do not necessarily correlate with animals raised for a specific quality of meat. These are two different parameters and the latter requires greater investment in time and effort, which is justified only if there is an economic or social incentive. The desire to equate more flesh with better meat is based on the modern context, where these distinct aspects of husbandry are closely correlated.

In turn, this feeds into a desire to suggest meat cuts and, by proxy, reference status (Stokes 2000: 147). On what are these assumptions based? Are data describing demographic proportions of sex and age adequate? Considering that our methods amalgamate data from many individual animals, is it actually possible for us to note meat cuts? Divisions of the carcass certainly exist, based on natural points of weakness in the morphology of the animal. However, these are fundamentally different from the modern iteration of 'meat cuts'. This latter category is a form of industry standardisation, which has historically been defined by region (Gerrard 1979: 321–2; Swatland 2000). These two distinct aspects – both central to the transformation of flesh to meat – have become conflated when viewed from the archaeological context: observing carcass divisions but assigning standardised 'cuts'. Thus, notations that suggest meat cuts apparently derive from the contemporary context and represent a misappropriation of modern analogy.

For the archaeological context, in numerous cases the evidence supports a pattern of husbandry centred on meat production, often as a consequence of maintaining large herds for traction (O'Connor 2014). This does not tell us about the type of meat in question. We may have indirect evidence of the quality of flesh based on age; older animals are likely to have more uniform musculature. However, we should avoid trying to infer status from a perceived notion of 'good meat' or 'better cuts' (Stokes 2000). There are areas of the carcass with more flesh, but that does not necessarily equate to better quality meat. In short, zooarchaeologists would benefit from better integration of faunal datasets, quantitative and qualitative, in order to provide a more representative picture of animal husbandry from the past.

Finally, there is also the need to position the social context. This is where the work by Stokes, Sykes, and numerous other authors is driving faunal studies to tackle social dimensions. Any concern about the potential validity of the outcomes from these studies rests in how the methods to study cut marks actually help to reveal social features. Returning again to the example of ethnoarchaeological research with the Maasai (Seetah in prep b), the relationship between Maasai pastoralism and meat consumption offers an ideal example for why there is a need to exercise caution.

For Maasai, meat is the most important food source, alongside blood and milk, and is considered essential for well-being. Decision-making processes for how an animal is butchered are based on culinary, medicinal, and most importantly, sharing drivers. There are strong connections between communication of specific ideologies and consumption. Furthermore, there is a well-structured system in place to retain and maintain these social hierarchies, which begins with instruction in the proper processes of butchery and subsequent sharing. These communally defined mechanisms of food sharing garner social capital in numerous ways and help retain strict boundaries between individuals at the level of the household, i.e., within the *boma*, and at a community level. In Maasai society, the part of the carcass that you eat effectively exemplifies who you are and illustrates an excellent example of 'social' and hierarchical eating – a concept noted by Stokes from Roman South Shields. Table 8.1 illustrates the main details of how a goat would be shared in Maasai society. Divisions are allocated based on the gender and age of the individual (i.e., the 'age set' to which they belong).

TABLE 8.1 *Social and hierarchical food-sharing drivers*

Carcass portion	Recipient
Head	Head of the family/older male from the 'senior age' set of the group
Neck	Meat (only) is always provided for strangers/guests. Family providing the animal will use the bone for soup.
Shoulders	Generally given to women of 'middle age' (35–50 years old); however, this portion is shared with some flexibility and alternatively can be given to warriors.
Back and saddle	These two portions are provided to young, uncircumcised girls who will share between themselves only. This will include all young girls in this set from the local *boma* (homestead), as well as neighbouring *bomas*.
Ribs	Considered a special portion; at least one part will be given to the senior men; alternatively, the ribs will be given to special guests.
Breast and belly	Young, uncircumcised boys will receive this portion, often cooking it themselves to be shared by other members of the age set.

Carcass portion	Recipient
Fore legs	Given to women of 'middle age'
Back legs	One back leg will be given to the women of 'young age' (circumcised, 20–35 years old). The women will cook their portion as they see fit. The other leg is given to the senior man; the bone will be used for soup and meat for steaks.
Sacral region	Reserved for 'older' women (above 50 years old).
Body 'fatty layer'	Always given to the owner of the animal.
Liver	In this case (goat) is given to senior elders; had the carcass been a sheep the whole liver would be given to the women present.
Lungs and heart	Young, uncircumcised boys
Intestines	Given to the 'young' women
Kidneys	This portion is reserved solely for the person carrying out the butchery, although that person may relinquish it. This portion is considered safe to be eaten raw and is often consumed during the actual butchery process.

Sharing follows a rigid hierarchy; apportionment of the back, saddle, head, and neck steaks (not the neck bone, however) is strictly adhered to. However, there is a degree of flexibility. Negotiation takes place depending on why the animal has been slaughtered. An example of this is noted with the allocation of the shoulder. This portion can be given to members of the warrior set (young men tasked with defending the community). However, if the animal has been killed because a birth has taken place and the mother requires stronger nutrition, then the shoulders are given to the middle-age women (between 35 and 50 years). This set will be responsible for cooking and preparing the meal and administering to the individual(s) in need.

Maasai generally consume with other members of their respective age set, and particularly during a feast will disperse into groups. The bones are discarded around these respective groupings. If a waste assemblage from such an event is analysed, the age and sex profile would almost certainly be interpreted as a profile of husbandry that indicated animals were raised for meat. However, this is not why Maasai keep and raise cattle, sheep, or goats. The social dimensions, revolving around sharing of *meat*, dowry gifts, and notions of wealth tied to animal bodies, are infinitely more important than the nutritional value of the animals' flesh.

On Butchery and Butchering

It is easy to comprehend why analysts encounter difficulties regarding the material itself, animal flesh. Zooarchaeologists are experts in animal bone and less likely to be as familiar with soft tissue. Moreover, as Chapters 5 and 6 illustrate, the techniques and tools involved in butchery form complex topics of research in their own right.

Thus, identifying misconceptions that have arisen as a consequence of unfamiliarity with flesh serves as a useful starting point for bridging the modern and archaeological contexts. Maltby (1979: 38; 1984: 128) indicates that large numbers of skull fragments from cattle found on Roman sites are attributable to 'smashing the skull for brain removal'.[3] Logically, it would follow that the soft tissue of the brain would also be smashed if the cranium were treated in this way, therefore hampering the recovery of the desired tissue. Alternatively, this method has been countered by suggestions that the brain was removed following chops to the suture lines (the meeting point of the various bones of the cranium) (Thawley 1982: 219). This is an unlikely scenario given the stages involved. To remove the brain from a cow's skull along the sutures, as Thawley supposes, a practitioner would first need to skin the head, remove the muscle around the maxilla and cranium to reveal the sutures, and then chop into these lines. In older cattle, commonly found on Romano-British sites (Albarella et al. 2008), these are likely to have been obliterated through growth and age. The simplest way to remove the brains is to deliver a single chop through the middle of the skull once it has been skinned. As there is little meat around the cranium, there is no cushioning of the blow, and the full force of the chop is directed cleanly through the thin bone of the skull, past the brain, and (provided sufficient force is used) through the maxilla – leaving the brain readily available (the method reported for game in early work by Guilday et al. 1962: 77). An alternative would be to deliver the chop from the palate to the cranium. In each case, either the maxilla or the frontal bone provides a flat surface to rest the head and stabilise the chopping process.

Two tangential points are worth noting here. The first example above highlights why the language used to describe butchery would benefit from greater precision. Improved terminology could also help to better acknowledge the skills and experience of practitioners. In response to Thawley's conclusions, the alternatives I propose are presented as potentially more plausible techniques for brain removal. My intention here is not to be rigid. A range of techniques could be used and there is no a priori reason as to why the method suggested by Thawley is not possible; it is only improbable for the reasons outlined.

Other examples worth noting include the observation that 'fresh long bone is extremely strong and resists breaking unless subject to severe stress or to attack by an implement designed for the purpose' (Done 1986: 144). While Done identified the need for specialised implements, the material – fresh bone – is misrepresented. Fresh long bone is easily marked, quite brittle, and easily fragmented if it is chopped in the right place for breakage. In fact, in modern Western butchery, the aim is often to avoid breaking the bone as it invariably splinters, and it is for this reason that the saw, and recently the electric band-saw, have gained favour. Furthermore, this example once again highlights the need for better descriptive terminology; 'attack' hardly seems appropriate to describe skilled practice.

Misconceptions also relate to how the implements themselves might have been utilised. Maltby, reporting from Wantage, Oxfordshire, suggests that 'dismemberment and sometimes filleting using a cleaver or heavy blade became the normal methods of butchery on most Romano-British sites' (Maltby 1996a: 159). While dismemberment can and does occur using a cleaver, filleting would be impractical using a large and relatively heavy implement. The basic requirement of a filleting knife is to be sharp, small, thin, and light, with a relatively flexible blade. A cleaver does not possess these qualities.

Marks are often attributed to cleavers when in fact they were more likely to have been made with a blade.[4] In part, this is because a sharp knife, even a small one, may leave a mark that can be misread as a large blade. Thus, we need to exercise caution and not assume that a large utensil is the only implement that can leave a deep blade mark. Other examples of misrepresentation and disconnect between tool and task include the reporting of chop marks on the mandibles, in the region of the diastema, as indicative of butchery for removal of the tongue (Done 1986: 144). The tongue inserts into the throat, and specifically attaches to the hyoid. The diastema forms the space between the incisors and the molars – in other words, at the opposite end of the mandible to where the tongue inserts. Further, there would be no requirement to chop the tongue to remove it; this is easily achieved using a blade.

The saw seems to be the most misunderstood tool. Maltby proposes that 'cattle carcasses on urban sites appear to have been more intensively processed for their meat and marrow, mainly through use of heavy implements such as the cleaver or saw' (1993a: 319). It has also been suggested that saws may have been used to fillet meat from the bone due to scoop marks seen on the shaft (Maltby 1993a: 15 [microfiche]). While the cleaver can certainly be termed a heavy implement, the saw is in fact prone to breakage and blunting, as the teeth by necessity have to be small and compact to deal with dense bone (compare a saw used for metal cutting, which needs to have small, closely spaced teeth, with a wood saw, which has larger, well-spaced teeth). A saw cannot be used as a filleting tool as the teeth would quickly become clogged with soft tissue, particularly if used near tendons and ligaments. The scoop marks are therefore more likely to have been caused by a slicing motion with a knife, utilised close to the bone, and resulting in flakes of bone being pared off.

Misrepresentation and Misinterpretation from the Butchery Record

The cases above are sufficient to illustrate how misunderstandings can lead to misrepresentation. An account of Roman butchery reports:

> Modern butchery practice is to split a hung carcass down the axis, then quarter it, removing the limbs in sections. The Romans placed the whole carcass flat, then removed each limb in preparation to being further divided. Heavy knives and cleavers would make short work of this. There does not seem to have been any boning of the joint. Marks on bones suggest meat was scraped off, perhaps even shredded. At Neatham, in Hampshire, badly chopped shoulder blades indicate heavy, unskilled butchery.
>
> (Alcock 1996: 23)

The misconceptions represented in this quote arise as a result of inaccurate reports from archaeological finds. There is little evidence to support or reject whether Romano-British meat processors placed carcasses flat for dismemberment, and in fact cuts on metapodial bones from Roman sites would suggest animals were hung (Hambleton & Maltby 2009). Logically, there would have been some deboning of joints to fit the meat into pots for cooking; scraping and or shredding meat off the bone would not only be impractical, it would also result in meat that is unlikely to be saleable. 'Badly chopped shoulder blades' are more likely a result of specialist butchery to trim the scapula for smoking or brining (Dobney et al. 1996). Finally, suggestions of 'unskilled' butchery do not hold true when we consider the limited evidence from archaeological contexts to infer levels of expertise.

The final point is particularly pervasive in zooarchaeological literature and is equally misleading. In 1989, Serjeantson suggested that the nature of cut marks recovered from Romano-British assemblages were indicative of crude and unskilled practices (1989: 5). These observations need to be taken within the context of the types of marks Serjeantson was describing, the relatively nascent state of Romano-British zooarchaeology, and the available methods for cut mark analysis at that time. The marks themselves were idiosyncratic. They were created with large blades (or more likely cleavers) and indicated a system of butchery that revolved around chopping.

When these types of signatures were contrasted with cut marks from British Iron Age bone assemblages, the differences were striking. Although there is no a priori reason to think that using a cleaver requires less skill than using a knife, one can appreciate that large, irregular marks, when compared with fine knife marks, might be thought of as crude. This is not the case. Using a cleaver requires a different type of skill, closer attention to the cutting process, and careful hand–eye coordination. Much greater forces are applied during chopping activity. A butcher may suffer a slicing cut to the hand or fingers when using a knife, but these are usually minor; a misplaced chop can result in a lost digit – a far more serious injury.

While early explanations of Romano-British butchery may have been conservative, without capturing the nuances of the craft, they are understandable given the context. Unfortunately, despite numerous important studies that have illustrated the complex nature of Romano-British provisioning and the role that the skill of the butcher played in developing the trade (Maltby 1984), such statements continue to pervade the literature. Furthermore, these comments are not restricted to the Romano-British case. Describing the cut marks found on cattle skulls from a Late Bronze Age/Middle Iron Age ditch site at Battlesbury Bowl, Hambleton writes:

> Repeated marks in the same location could result from inexperienced or over-enthusiastic knife work, or simply be the preferred method of the butcher. Equally, repeated knife cuts could indicate particular care in removing any adhering soft tissue, perhaps more so than would be required during routine skinning and filleting. (Hambleton 2013: 488)

The author clearly identifies and emphasises that a range of possibilities is likely from the geometric cluster of marks. Hambleton goes on to make the point that the marks are specifically for cleaning, above and beyond basic skinning, in order to prepare the skull for display. This commendable level of detail expresses the level of nuanced interpretation that is possible from butchery data. However, there seems a default predilection to suggest inexperience or over-enthusiasm, which one might read as contemporary synonyms for crude and unskilled.

An example from a recent and important book on social zooarchaeology illustrates how the limitations imposed by current recording methods continue to hamper our ability to interpret from the butchery record. It also demonstrates that these inadequacies are effectively becoming entrenched, which seems to have led to a widespread – and negative – view of the butcher in antiquity. Sykes (2014) triangulates between her own experiences of slaughtering and butchery of two chickens and research from Iron Age, Anglo–Saxon, Roman, and medieval bone assemblages from Britain to draw the following conclusions:

> When we killed and butchered our cockerels Gunter and Nightshade, we no doubt left slightly different traces on their skeletons: the two events were different reflecting the different relationships we had with these two individuals. In both cases the butchery was done carefully and respectfully because we actually liked our cockerels. For this reason, we did not pick up a cleaver and rapidly chop them into evenly-sided portions; aside from the fact it would have caused bone splinters,[5] it would have felt morally inappropriate. No such consideration was afforded the animals that made their way to market in Roman and later medieval towns. Butchers presumably had little time to build a relationship with the animals they dispatched and any hint of character could quickly be erased by transforming the individual into anonymous and standardized cuts of meat: chop, chop, chop. The Roman and later medieval butchery marks shown in Figure 1.5b scream of a situation where animals are viewed as commodities rather than individuals, something that is less apparent from the Iron Age and Anglo–Saxon butchery patterns (Figure 1.5a) and, I would suggest, this tells us something about the closer human–animal relationships experienced in these period. (Sykes 2014: 15–16)

This paragraph captures a specific set of issues that should merit the close attention of those interested in accurate representation from the butchery record. Equating modern household to large-scale archaeological processing, or the butchery of chickens to cattle, is not a commensurate comparison. These examples are not appropriate bases from which to infer broader orientations of respect or morality during the Roman or later medieval setting. Why would it be any less moral to butcher with a knife, as opposed to a cleaver? Is careful butchery more moral? Or is rapidity of processing the issue? Did butchers per se, whether in the Iron Age or otherwise, ever have time to 'build a relationship with the animals they dispatched'?

Two conclusions emerge from this discussion. The first is that a comparison between modern trade butchery and urban/military Roman or commercial medieval sites would have been a more appropriate evaluation and would probably have revealed closer correlations (Seetah 2004) (as would a comparison between contemporary chicken butchery and rural or household activity in the Roman or medieval periods). The second is that the legacy of a negative view of the butcher, based on conclusions drawn from material that was not well understood, permeates contemporary research. These points highlight the pressing need to situate craft and craftspeople and to present more egalitarian and better-informed views of those who created the assemblages we depend on to study the past. Equally important, these incommensurate comparisons do not make the best use of the data in formulating interpretation.

Sykes depicts a scapula and a femur in her book; in Fig. 8.2, I use a humerus to depict similar patterning. Based on large numbers of bones that are regularly recorded from later period sites, a perceived idea has developed suggestive of rates of slaughter and butchery: many animals processed in a short space of time. Unfortunately, this is an example of correlation, not causality. It is possible to generalise and suggest many more animals were processed if we compare the Iron Age with the Roman periods, for example. However, our ability to estimate how many animals were actually processed on a daily, weekly, or monthly basis is limited. Moving from the faunal data to the butchery record specifically, the oft-used representation of the cut marks characterised by Fig. 8.2 is problematic. By attempting to categorise all the marks from an assemblage on a single diagram, we are conflating evidence from a range of individual animals and creating an artefact of those data that we use to suggest a pattern of butchering behaviour.

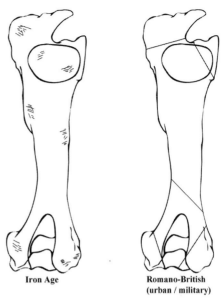

Iron Age

Romano-British (urban / military)

FIG. 8.2 Typical arrangement of cut mark variability and type of cuts between Iron Age and Roman British practices.

In effect, this amalgam of data is utilised to posit an occurrence that has never existed. Part of the problem with this type of illustration is that we are giving the impression that this depiction characterises what is taking place on one animal, which is not the case unless we find all these marks on a single individual. The more significant concern is that these are accepted as factual when they are actually illustrative of a generalised pattern. Representations of butchery patterns produced from amalgamated data need to take second place to more direct interpretation. These representations do not offer a rigorous basis from which to organise data or a point from which to build inference. The approach adopted by Sykes is commonly used, but ultimately homogenises data, drawing us away from the individuality of different butchery episodes. Sykes's point is an important one, aiming to position the role of the animal within a wider socioeconomic context. However, to suggest that Roman and later medieval practices 'scream of a situation where animals are viewed as commodities rather than individuals' is to overstate the case, particularly when no individual animal actually underwent the processing that the diagram characterises. This does not, however, invalidate the use of such general patterns. Clearly, in this case, it serves as an important mechanism showcasing diachronic variation.

Sykes suggests that the butchery record shows evidence of an 'overhaul of human–animal relationship' (Sykes 2014: 15). This is precisely the type of inference that the butchery record can reveal, and the author is to be commended for drawing our attention to this conclusion. I have suggested that cut marks from urban Romano-British assemblages, for example, could be taken as an indication of increased commodification of cattle (Seetah 2005: 6). My own conclusions were based on a comparison between Roman rural and urban sites and associated technology. While Sykes's research illustrates the potential of the butchery record, it also highlights how deficiencies in the way we record and interpret butchery can lead to problems when generalising from the gathered data.

CONCLUSIONS

New and exciting developments have recently been made in zooarchaeology. In particular, opportunities in applied zooarchaeology (Wolverton & Lyman 2012) and advancements made in molecular analysis of animal bone (Guiry et al. 2015; Hagelberg et al. 2015), animal palaeopathology (Vann & Thompson 2006; Bartosiewicz 2008b), bone working (Choyke & Schibler 2007; Choyke & O'Connor 2013), and the theory of social zooarchaeology (Russell 2012) are redefining the subdiscipline. Butchery, a vital aspect of faunal assemblages, has not been swept along with this impressive wave of new endeavour.

Throughout this book, I make a concerted effort to illustrate the depth to which butchery pervades society, and the complexity of the practice. This complexity is both the main strength of butchery and its Achilles heel. The opportunities for contributing to archaeological interpretation are palpable. We can see the actions that are trapped within this proverbial amber. Alas, for the analyst, each cut mark is quite literally a unique entity that captures physical and cognitive characteristics of the practitioner, as well as properties of the tools used for butchery, driven by a host of external factors. How can we unravel such a tangled web of information? First, we need to address past concerns and inherited methodological limitations. The structural frameworks within which we work inhibit efforts toward standardization. Second, and more problematic for the analysts, misperception of the subject has meant that it has become symptomatic to qualify the data by its quantity.

This is all the more unfortunate when we consider the main point of Chapter 3: from the perspective of *chaîne opératoire*, numerous new guiding principles are being developed that would greatly enhance our ability to study the butchery record *as practice*. Unfortunately, recent scholarship seems to be moving further down the path of emphasising the mark. Techniques that depend on GIS (Abe et al. 2002; Orton 2010) and template recording systems (Popkin 2005), while serving as an excellent means of representing the marks – which the respective authors make clear is the main aim of these methods – do not actually address the underlying issue of how to better record process. Such methods have not been widely adopted.

The solution to some of the issues identified above is relatively straightforward. We need to relinquish the idea that this dataset can be recorded with absolute accuracy; it simply cannot. We will never be able to record or interpret all that was taking place during the process of butchery, as well as the drivers involved. We then need a recording system that actually synthesises the data that are captured within the butchery record: details of the mark, the tools, and the function. This may seem obvious, but the underlying problem is more deeply seated. As this chapter illustrates, the root of current limitations lies in the way we think about butchery and cut marks.

We have a precedent for recording in specific ways that has had negative ramifications for how we study butchering. On a practical level, the promise of butchery is unlikely to be realised unless we are willing to move away from coding systems that obscure the data and do not actually record the details we are interested to recover. We need to reconsider all aspects of our analytical process, and how we then extrapolate our interpretation. Zooarchaeologists record copious numbers of marks – tens of thousands – but report on only a small suite of those marks that tell us about skinning or meat removal, where these were found on the carcass, and what they might mean, e.g., whether there was hanging or not. Despite caveats that we should not expect to recover all evidence from the butchery record, our current approach is actually recording redundancy. If we are using only a proportion of the butchery record, then we are failing in our endeavour to use animal bone to study the past. Analysts need to be adept at recognising the nuances of cut marks, but then move beyond this to assess the activity involved in butchery and the depth of socioeconomic drivers for butchering.

In addition, experimental replications – a cornerstone of this specific topic – need to consider the knowledge that the butcher likely possessed, examining the evidence for variations in levels of skill or quality of work, and attempt to situate these dimensions. We would never assume that an unskilled person could throw a pot, create a hand-axe, or fashion a glass object without prior experience. One might also anticipate that a degree of control would be maintained over the raw materials used. In contrast, much analysis seems to assume that butchery can be undertaken with minimal expertise and that all meat is equal, irrespective of where it originates on the carcass. Neither assumption has merit. There is also a need to take the operational sequence of butchering into account and control for variation in skeletal morphology (Machin et al. 2005). Once again, the principles of *chaîne opératoire* can serve as a guide as we develop our approach along these lines. Above all, first we have to acknowledge (contrary to Sykes, 2014: 16) that the sum of the practitioners' knowledge amounts too much more than 'chop, chop, chop'!

CHAPTER NINE

HARNESSING THE POWER OF THE CUT MARK RECORD

INTRODUCTION

Part I has served to deepen our understanding of butchery, using modern analogy from industrialised and non-industrialised settings to position the subject matter within wider societal, technological, and practical contexts, identifying the routes to specialisation in practice. By appreciating the craftsperson, the craft, and the knowledge, we are prompted to situate *the butcher*. Analysing butchery through the lens of performance, as a *chaîne opératoire* of production, allows us to better understand the nuances of butchery practice, including its phenomenological attributes. Improving and enhancing current approaches to recording marks rests on examining more closely the interplay between the agency and physicality of the human actor, the animal's body, and the technology that serves as an interface. This chapter takes up these perspectives to respond to the inherited limitations identified in Chapter 2 and interpretational concerns noted in Chapter 8.

A more comprehensive overhaul of butchery in the future may, for example, include creative new studies that are focused on developing an understanding of the ideology that drives butchery practice or take an anthropological view of the relationship between craft specialism and butcher castes (for similar work on pottery, see Roux 1989). However, as part of methodological treatment of butchery, one development to prioritise is enhancing our recording protocols to capture 'process'.

Efficiency (Spiess 1979) and in particular transport (Binford 1978; Lyman 1994: 299) have served as the dominant conceptual frameworks from which to assess butchery. An emphasis on process, in contrast to location and typology of the marks, builds on the more holistic conceptualisation of butchery as activity, craft, and practice. Process describes the fact that the activity that cut marks represent is more complex than the relationship between the knife and the modification on a bone. A useful analogy would be the way a carpenter works wood, leaving residues in the form of tool marks, a signature of production (see Bealer 1972). For the woodworker, there is often a need to remove the marks through sanding and polishing to realise the finished product. For the marks created by ancient butchers, a similar process happens through taphonomy; however, cut marks are more indelible as they are rarely deliberately removed unless this forms part of subsequent working.

Butchery data incorporate subsets of different types of evidence; furthermore, cut marks are a means to an end. Both of these points necessitate an applied methodology. The analyst is assessing variation in all of the following: geometry and morphology of cut marks, implements, and activity patterns that are driven by agency. This trinity should intuitively lead us to consider the recording of cut marks as a tripartite, multilateral task (Seetah in prep a).

The approach I will outline in this chapter rests on a combination of taphonomic and cognitive data, harnessing the strengths of both perspectives and integrating the flexibility of template recording. Recording geometry and frequency of cut marks needs to be corroborated and enhanced with evidence from the tools and a better understanding of the functional characteristics of the marks. This will assist with the interpretation of what the morphology and arrangement of the marks signify. The aim of the approach, at a disciplinary level, is to provide the foundation for future standardisation of cut mark analysis and interpretation. This includes standardisation of terminology used to describe the recorded butchery activity and tools. For the present case, this is built into the method by adopting terminology from modern butchery textbooks and terms already used regularly in the archaeological literature but restricted to defined meanings. Finally, by using the same structural framework as contemporary recording protocols (i.e., computer-based systems) but branching off in several new directions, the approach is easy to integrate and retains the strengths of modern databases, complemented by recording on templates.

COMBINING QUANTITATIVE AND QUALITATIVE ASSESSMENT OF CUT MARKS

In summary, the goals of this approach are to facilitate the analysts' ability to do the following:

1. Identify the implements that were most likely to have been employed during carcass dismemberment

2. Clarify the operational processes that created the cut marks
3. Provide ways to integrate, assess, and visualise the data recovered from tapho-nomic markers, implement signatures, and functional datasets.

Despite the long-held admonition against overemphasis of the mark (Dobney et al. 1996), when it comes to metal-tool butchery, there is still a need to improve our ability to differentiate between tool mark signatures. This is necessary to determine the implement used during butchery, as illustrated by those working on lithic butchery or those distinguishing between lithics and metal (see Fisher 1995; Greenfield 1999). For later periods, we have yet to establish clear categories of metal tools. There has been a tendency to conflate action and implement. As such, recording a 'slice' also implies the tool, a blade. However, different types of slice marks exist (Greenfield 2006), as do different types of blades (summarised below, from Seetah in prep a). Equally important, the type of detailed assessment of marks characterised by Greenfield (1999, 2000; see also Bello & Soligo 2008) is essential to assess some of the parameters that reveal *process*. Evaluating directionality, for example, has an established precedent for lithic-tool butchery (Bromage & Boyds 1984) yet has had little influence on metal-tool butchery or, indeed, on cut mark recording protocols. As the following outlines, determining the direction of marks and positioning the butcher and carcass in relation to one another are critical to any attempt to situate how technology mediates the transformation of carcass from flesh to meat, and the attendant benefits this may have for interpretation of the wider human–animal relationship.

At a more fundamental level, the approach tackles the issue of 'coding' in a direct way. Again, referring back to Chapter 2, a lack of standardisation in zooarchaeology has been lamented and is unlikely to change soon. I adopt a more pragmatic approach.

In general, coding systems in zooarchaeology, as with other subdisciplines in archaeology, are a highly usable and essential means of cataloguing the large datasets used by material analysts. Systems use either numeric or alphabetic codes. Numeric codes are favoured in those situations where many species are encountered in an assemblage. Thus, for North American contexts, numbers are used to reference a species: '163' equates to cattle. For historic periods in the United Kingdom, where the species diversity is lower, a letter code may be used: 'COW'.

By this stage in the book, the reader will appreciate the complexity of trying to record cut marks to accurately reflect the full breadth of evidence that is contained in this dataset. The reader will also appreciate why 'coding' has been part of a progression to record butchery, as individual analysts attempt to capture the details of the cut mark. Figure 9.1 illustrates the type of schema used to record cut mark location and orientation, based on Binford's work with the Nunamiut (where he depicts the marks on a deer skull; here I have used a cow) and essentially grounded on a range of earlier studies (Binford 1981: 131–42, table 4.04). Lyman (1987) and Maltby (1989: 92–102, see tables 2–14) have developed similar protocols. These annotated codes often rest on a basic premise that attempts to combine the data from standard recording of faunal data with butchery evidence. Thus, the 'cut mark record' is incorporated into the 'faunal record' and includes evidence of the bone itself as part of the 'code'. In the example illustrated in Fig. 9.1, S1, in this case a code for the

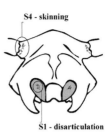

FIG. 9.1 Examples of cut mark coding.

skull, indicates 'transverse cut on occipital condyles'. The same code, S1, can be applied to the scapula, but then refers to 'marks along inferior border of condyle and/or at origin of triceps brachia'; in both cases the activity denoted is dismemberment. For the skull there are seven such categories, S1–S7, and four for the scapula, S1–S4.

Such schemes have a number of problems that have tended to be exacerbated with time. To recap an argument made in Chapter 2, codes effectively constrain the analyst to fitting the observed cut mark into a specific category. In Binford's system there are 108 categories to describe all the butchery that could occur over the carcass. This is insufficient and unfortunately redundant. The overall butchery process is restricted by the animal's morphology, and clusters will occur, for example, due to skinning. However, cut marks can conceivably occur in an infinite set of arrangements, locations, and orientations. The marks are also mediated through culture and technology. It is for this reason that recording systems over time have consistently had to add categories, which ultimately renders them unwieldy, specialised, and non-transferable. This type of coding is also difficult to integrate into a database format, as invariably there will be some variation from the stated norm. Thus, such coding is undertaken as a separate task, usually in the form of an annotated text file and often complemented with templates. In practice, these supplementary files, containing a wealth of detail, rarely see the light of day, as they are inherently unstandardised.

As the above should make clear, coding is an essential part of populating zooarchaeological datasets. However, this has been a limiting factor for butchery recording. What is needed is a transition to a more descriptive methodology, but one that can still be integrated into contemporary database systems. To promote and facilitate adoption by the wider zooarchaeological community, the system discussed below, or its constituent parts, can be streamlined into existing recording protocols and applied across a wide range of periods and geographic locations where metal tools have been employed for butchery. A comprehensive discussion is presented in Seetah (in prep a) and summarised below.

The recording protocol is optimised for ease of input as part of a computer-ised database. This facilitates not only the recording phase itself but also subsequent analysis and the recognition of patterns. Database recording is almost universal in zooarchaeological analysis. Integrating a 'butchery sub-form', similar to subforms used to record tooth wear, for example, does not therefore represent a major departure from schemes with which the majority of zooarchaeologists are familiar. The recording form was developed from Bournemouth University's Faunal Recording system and modified to accom-modate butchery parameters through a series of drop-down menus. The system is adapted to accommodate cut mark variability and takes into account the fact that, on the same bone, there will often be multiple occurrences of the same mark or different types of marks that could themselves each occur in multiples. This same principle extends to the recording of implement signa-tures and functional characteristics of the the butchery activity.

The essential innovation is a simple and intuitive one: rather than attempting to fit the cut mark into a preconceived and, by design, structurally rigid framework, the recording framework itself is modelled around the dataset, to accommodate different aspects of the cut mark. The next section summarises the parameters that form the basis of the recording protocol. These parameters include butchery-specific[1] components that are tailored to reveal process, encouraging the analyst to consider the relative position of the butcher and to interpret those aspects that are intrinsic to butchery, but not immedi-ately obvious from the cut mark itself. I then present a case study to illustrate how these parameters assist with interpretation.

Butchery Parameters Recorded

Observational characteristics (*characteristics based on direct observation*)

- Butchery location: refers to a generalised notation to the area on each bone where cut marks are observed. This is specified as the joint facets only, the cortical surface only, or both. A further criterion under this category is 'margins', which indicates that marks are observed on or around the margins of the joint facets themselves but not directly on the joint or the bone surface. This provides an initial indication as to whether there was evidence for disarticulation (i.e., from marks on the joints), meat removal (i.e., presence of cuts on the surface), or a mixture of activities.
- Surface location: this indicates the physical location of the cut mark, with the *bone placed in correct anatomical position*. This parameter is used as a means of establishing a standard reference point based on the anatomy of the animal. In turn, 'surface location' provides a means of relating the cut mark to other landmarks on that element and other body parts. This parameter can then be

compared with subsequent records that indicate whether the butchery took place with the element out of correct anatomical position.

- Direction of mark: the trajectory of the mark *relative to correct anatomical position*. For example, was the cut mark created from the anterior to posterior facets of the bone? This is again compared with subsequent records as a means of deducing if the element (and therefore carcass part) showed evidence of being repositioned during the butchery process.
- Multiple occurrences: indicates whether there is more than one occurrence of any cut mark category (thus, taking cut mark variability into account).
- Depth: the depth to which the cut mark penetrated into the bone. Four categories are used: shallow, moderate, deep, and cut through.
- Implements:[2] which implement was used in the creation of the cut mark? Implements can be assigned to one of four categories: cleaver (includes Romano–British cleaver), large blade, fine blade, and saw.

Interpretational characteristics (*characteristics deduced directly from the mark*)

- Type of mark: aside from the usual chop, slice, fine slice categories, a number of additional types are included:
  - Point insertion, denoting that only the tip of the blade was being used
  - Blade insertion, indicating that a portion of the blade was being used for the cut, but without the blade being drawn along the bone, as might be noted with a slice
  - Scoop, implying that a scalloped piece of bone was removed following meat removal from the bone
  - Knick, indicating a small action created by a flick of the wrist and not resulting in a true slice.
- Pre/post disarticulation: this parameter provides a means of noting whether there are indications of when the cut mark was created, but only in relation to gross disarticulation. For example, were the marks created after the carcass was halved?
- Bone position: the position of the carcass portion during butchery. For example, using the characteristics of the mark, such as directionality, can the analyst deduce whether the carcass was lying on the ground or suspended?
- Direction of cut: as distinct from the 'direction of the mark' (an observed and absolute category), the direction of the cut is speculated in relation *to the practitioner* during butchery.
- Function: what was the underlying activity as indicated by the cut mark? For example, was the mark for skinning?

The expectation in recording the above characteristics is not that all the criteria will be fulfilled for every incident of butchery, although that is the ideal situation and often the case. Rather, each parameter contributes a small component to the overall view of the type of butchery activity under investigation. The above is then integrated into the standard zooarchaeological assessment of the bone assemblage. From the above, two categories require further explanation as they rely on features I have integrated from modern analogues and include specific details that are not universally recorded: 'type of mark' and 'function'.

Type of Mark: Implement Signatures and Corresponding Activity

Chopping Activity

Chop marks, created specifically with cleavers, are characterised by whether they appear on the cortical surface or in/around the cancellous bone. When present on the cortical surface, there is a tendency for the chopping action to leave a smooth surface at the point of impact. There is rarely a corresponding smooth surface at the opposing end of the chop, as the bone is invariably fragmented (Fig. 9.2); it is the transferred force of the blow through the bone that results in the bone breaking. This type of smooth surface on cortical bone is rarely evident when chopping into elements composed of cancellous bone, such as the epiphyses of long bones, as there is rarely enough cortical bone to show the smooth cut surface. Chopping into cancellous bone tends to be observed through cleanly fragmented and sharp angular 'facets' on the cut surface; if the blade is adequately sharp, such faceted marks can be distinct.

Slice

Slicing cuts are a ubiquitous form of butchery. This mark is characterised by a delineated striation along the bone, with a 'V' cross-section, although the length of the cut can be relatively short and compact. The depth of the mark, its length, width, placement, and number of occurrences at any single location on the bone, as well as portion of the implement used (i.e. the point or the blade itself), all show variability and are indicative of different activity patterns and butchery techniques. I have distinguished these into the following categories: fine slice, point insertions, and blade insertions, which can be used to gain greater depth of interpretation if these distinctions can be made. All of these subcategories are still created by the same implement; however, they are distinctive cut mark signatures using different parts of the knife blade and should be

FIG. 9.2 Chop marks on proximal femur.

distinguished from each other where possible. In this way, the analyst is afforded a more in-depth appraisal of activity and butchery technique.

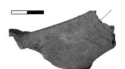

FIG. 9.3 Point insertions (arrowed) on a thoracic vertebra.

Fine Slice

A fine slice is distinguished from a slice mark by the fact that it is made with a sharp bladed implement with a thinner cross section than a typical 'slice' mark. Similar characteristics evident with a 'slice' are also typical of a 'fine slice', although the mark will be shallower and thinner. A fine slice implies a variation in tool type. For example, a skinning knife may have a finer blade than a knife used for normal slicing; construction of the implement, e.g., 'steeling' the cutting edge, can result in a slimmer and therefore sharper knife compared with an iron blade.

FIG. 9.4 Blade insertion.

Point Insertions

Point insertion marks indicate that only the tip of the blade was used during the butchery process (Fig. 9.3). Distinguishing this mark from other 'slice' and 'fine slice' marks depends on recognising distinctive insertion and exit points at the start and end of the cut mark. Furthermore, cut placement is important in recognising point insertions as these marks indicate use of the tip in a specific manner, during skinning for example, where the point of the blade is used to puncture the skin (inferring that the blade actually had a tip that could be used in this manner). Point insertions are not restricted to skinning. They may occur in other circumstances where only a small part of the blade is needed, for example, when filleting flesh from tight muscle attachments, or the tip of the blade can be inserted into a joint while it is still articulated.

Blade Insertion

This mark indicates a different portion of the knife being used, namely the blade. This type of mark is usually characterised by smooth entry and exit points of the blade. Rather than being confused with slice marks, this mark is more easily confused with chop marks, particularly where a large blade has been used. Figure 9.4 illustrates that the mark (arrowed) would appear to be a chop; however, this cut is in fact showing that a portion of the blade was inserted into the space between the occipital condyle and the atlas bone in order to decapitate the head.

Scoop

This distinctive mark denotes a blade insertion, but with clearer indications of functionality. In this instance the blade has been utilised along the length

FIG. 9.5 Scoop mark on long bone shaft.

of bone to remove small remnants of meat or to detach a portion of muscle from a particularly tight attachment to the bone. A characteristic scoop mark is left either if the blade is allowed to cut into the bone (Fig. 9.5) or if the blade encounters protruding bone architecture. As the force applied during this activity is usually minimal, the blade will 'stick' into the bone; in order to free the blade the butcher will flick and slightly rotate the wrist to generate a small amount of momentum. This results in a distinctive cut mark that is usually shallow and has a smooth entry point but a fractured exit point. The flick and rotation of the wrist breaks the bone, rather than pushing the blade through, although in some instances a complete flake of bone will be removed.

Knick

This type of mark is created in a similar manner to the scoop mark but is generally found in areas where the bone has complex architecture. Although similar to a scoop and often created for a parallel function, i.e., to remove meat, knick marks may also be present as a consequence of disarticulation. This mark is usually small and differs from the scoop and other blade marks in that it is characterised by an abrupt end point.

Sawing

For the periods under investigation, saw marks are more likely to be indicative of bone working and not butchery. Without exception, for the material studied for this book (see Chapter 10), they were noted only on metapodials and horn-cores. Furthermore, using contemporary practice as an analogy, archaeological saw marks are rarely noted on carcass parts that are sawn in a modern butchers' shop or abattoir. Saw marks are characterised by regular delineations on the surface of the cut itself. Furthermore, a saw will not fracture the bone as the 'cutting' (sawing in this case) is completed; therefore, the cut surface will demonstrate the striations through to completion.

Function

Skinning

This activity is characterised by cut placement focused around the lower limb extremities and the head, with an implement signature indicative of point insertions and fine slice marks. Knick marks can also indicate skinning activity. Within the operational sequence, skinning usually takes place directly after slaughter, but can occur before or after evisceration.

Disarticulation

This category denotes butchery for the removal of major limb elements. This is based on a combination of factors, most important of which was when, during the operational sequence, the particular butchery activity took place. Disarticulation marks are generally observed around joint articulations or on the joint surfaces. Alternatively, indications that chopping occurred along a group of elements, such as the spine, could also indicate disarticulation. Disarticulation can be carried out using either a cleaver or blade and is therefore not entirely dependent or characterised by one implement type.

Portioning/Jointing

This category indicates activity that would occur after gross disarticulation. It denotes the removal of muscles or groups of muscles from the bone, as well as sectioning of carcass parts, e.g., lower fore-limb from the upper fore-limb. As with gross disarticulation, a variety of tools can be used to accomplish the task, for example, a cleaver might be used to joint lower and upper fore-limbs. However, in general, as cutting of meat rather than fracturing bone is predominant, the principal tool used will be a blade, with subsequent cut marks generally indicative of point and blade insertions.

Paring/Meat Removal

This category is distinguished from portioning/jointing in that the marks are indicative of the removal of small remnants of flesh from the bone surface. This is commonly performed with a small sharp blade, as a larger blade would prove cumbersome and inadequate for the task; however, this does not rule out that larger blades may be used for this type of activity. Paring is characterised by scoop marks, with cut placement generally occurring on long bone shafts.

Filleting

This function results from the same underlying principle as paring/meat removal, namely to detach and exploit all possible meat resources. However, it is not always possible to use the blade of the knife against the bone to remove these final remnants. In this instance the tip of the knife is used to the same effect. Filleting also incorporates the removal of medium-sized portions of meat, again using the tip of the blade, but in a repetitive manner. This activity will be most commonly noted on awkwardly shaped bones (such as the scapula or groups of bones that are in close articulation, such as the vertebrae) or on parts of long bones that have particularly tight muscle attachments (such as the posterior tibia and the deep digital flexors and popliteus muscles).

Prep

This function indicates the use of a tool (not necessarily a blade; a sharp or curved cleaver can also be used) to expose a joint articulation for further disarticulation. This is effectively a precursor cut mark used to facilitate further cutting. It will usually form part of disarticulation activity, although it can be used during subsequent processing to allow a larger joint to be portioned or to section part of the carcass, such as the vertebrae.

Bone Breaking/Pot-Sizing

Bone breaking activity will usually occur at the end of the butchery sequence and may incorporate systematic fracturing of long bone shafts and epiphysis with a cleaver (Fig. 9.6a). However, more informative are indications of pot-sizing, which is characterised by chopping of larger bones into regularly sized units. Distinguishing between these two activities lies in the fact that 'pot-sized' bones will usually demonstrate more than one cut surface. This suggests a specific size requirement. Fig. 9.6b illustrates ribs that have chop

FIG. 9.6 (a) Bone breaking of distal femur, pot-sizing of (b) ribs and (c) skull.

marks at both ends and have been cut into regularly sized pieces; Fig. 9.6c shows a skull portion chopped on three planes resulting in a rectangular shaped piece. Both these examples indicate that the bone was cut and then repositioned and cut again in order to achieve the required size and shape.

EPISTEMOLOGY, TYPOLOGISING MARKS, AND RECORDING PROCESS

If notions of efficiency and transport represent the conceptual premise from which we currently assess the behaviour that drives butchery, then the animals' biology has historically served as the basis from which we record the marks. In turn, this is probably based on legacies inherited from the zoological foundations that underpin zooarchaeology. However, efficiency, transport, and morphology are not suitable starting points to record later period metal-tool butchery. Notions of efficiency and transport are significantly different compared with prehistoric contexts, becoming highly complex in historic periods and governed by intensifying scales of consumption. The constraints of the carcass are also less significant, as the tools themselves are designed to overcome this restriction.

Location is an important aspect in terms of recognising patterning, but marks can occur anywhere on the carcass; coding systems currently in use effectively require locational data to guide the cut mark record and this forces the analyst to emphasise biology, rather than 'archaeology'. An archaeological view effectively approaches the recording of marks from the perspective of the craft and realigns the spotlight to reflect on the agent, the animal's body relative to the agent, and the technology that is used to mediate activity. Viewed from the archaeological lens, we actually need to adopt a more 'typological' approach and to consider the cut mark as an archaeological entity – but not strictly taphonomic – that deserves a more systematic assessment. The approach outlined in this chapter still recognizes the importance of, and captures, geometric data, but achieves this as part of recording the butchery process.

Using the type of system characterised by the Bournemouth Faunal Recording Database allows for evidence of faunal biology and taphonomic, aging, sexing, metrical, and butchery data to be connected, rationalised, but also compartmentalised. Such systems facilitate the recording of a more complete dataset from the butchery record, including the actual number of repetitions of a mark at a given location, or variations in type of mark on a single bone. In addition, the approach I have outlined provides the analyst with a means to capture a wider configuration of geometric, typological, and sequential parameters, traditionally the strengths of template recording systems, but in a database format.

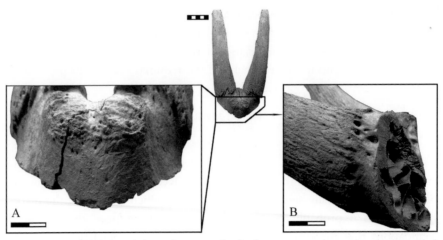

FIG. 9.7 Fine skinning and deep cleaver marks for horn-corn transportation (not removal per se). The assemblage ('Cēsis castle/town') that this cranial portion derives from dates to the period directly after 1214, from Cēsis, Latvia, when a castle was built by the Sword Brothers, and subsequently expanded by the Livonian Order from 1237. The castle remained in the possession of the Order until 1561. The town grew up close to the castle with a mixed German and indigenous population.

The following presents a case study illustrating how the steps through which the analyst can actually record process, emphasising the interpretational characteristics of this recording system. I then build this characterisation to situate this particular bone within the wider craft context.

The example in Fig. 9.7 shows clear evidence of tool diversification, including the presence of cleavers. The first set of marks indicates the use of fine blades for skinning (Fig. 9.7, inset A). Considering the size of the horn-cores and the repetitive nature of the marks, it is probable that the head was removed prior to skinning (post disarticulation). The skull was repositioned, as evidenced by the geometric location of each cut mark type; i.e., both skinning and chopping occur in multiple planes. The butcher, naturally, would have had to move the skull to complete each task; this would be easier to do with the head removed, particularly due to the presence of the horns.

Skinning was undertaken with the tip of a blade (point insertions). The skinning marks are travelling right to left and left to right (direction of the *mark*); however, relatively speaking, they are in fact directed toward the butcher, moving from 'front to back' (direction of the *cut*) – the butcher is drawing the blade toward their own body. The direction of the mark is varied because the head is being repositioned, not because the butcher is moving. Subsequent chopping to detach the horn-cores utilised a cleaver (Fig. 9.7, inset B) and required a blow to either side of the skull, resulting in the 'V' formation we observe on the frontal bone. In this instance, the bone position can be deduced by the fact that the striations created by the cleaver travel from lateral to medial (direction of the mark), indicating that the butchery was performed with the skull resting on a block. It would not be possible to chop cleanly through the skull unless it was positioned on a solid base. The resulting 'oblique' cuts are in fact top-down (direction of the cut) chopping blows. They appear oblique, as the skull is being held at an angle, using the horn-cores as a point of purchase.

At this stage, a reflexive moment helps to clarify details at a finer resolution. Taken together, the positioning, directionality, and various stages of repositioning that characterise the skinning and cleaver marks suggest that the cranium was likely held in the left hand, using the horns for purchase, and skinned/chopped with the right hand.

Finally, briefly emphasising the wider *chaîne opératoire*: these double horn-cores were found with regularity, indicating that they were probably transported with the cranium. The chop marks to detach this portion of the horn-core from the skull are too low to be an effective means of removing the horn. This indicates that a butcher carried out the preliminary processing to skin then detach and sequester the horn-cores (function), leaving the removal of the horn itself to an intermediary, either a horn breaker (Yeomans 2008) or a horn worker. Thus, these skull discard components show evidence of both butchery and preparation for horn working, situating different skills and craftspeople, and the interactions between them, within a wider economic and craft network.

CONCLUSIONS

An emphasis on process moves us, intuitively, away from protocols that are essentially constrained and repositions the analyst to record features of the mark that are descriptive of craft and human agency. At the same time, the approach outlined above retains the functional benefits of database recording schemes, and as such fits within the current ecosystem used to record bones. This latter point should promote adoption of a more standardised format for recording butchery (Maltby 2014). Furthermore, although the scheme I describe has been developed from and tailored for metal-tool butchery, the underlying principles and indeed the parameters themselves are applicable to the lithic-tool context. Ultimately, this recording system promotes a different view of the object, nesting the bone within the butchery ecosystem and shifting our view to one that more easily references the dynamic, biographical nature of animal remains. The next two chapters concentrate on the implementation of the methodological approach outlined in this chapter , demonstrating how the biographical, sequential nature of cut marks can be utilised to better assess the place of fauna in Roman and medieval Britain.

CHAPTER TEN

HUMANS, ANIMALS, AND THE BUTCHERS' CRAFT IN ROMAN AND MEDIEVAL BRITAIN

INTRODUCTION

The extended case study that follows operationalises some of the concepts and guiding theoretical principles introduced in Part I, based on the methods described in Chapter 9. The case study summarises data from six sites, which are used in Chapter 11 to examine broad trends in butchery practice from the Romano-British and medieval periods. Of these sites, one in particular, Cirencester, emerges as of special significance for our understanding of butchery in this region and time period. The resolution of data from this site was sufficient to be able to deduce discrete *chaîne opératoire* sequences for different parts of the animal. These sequences were based on accumulated results from numerous bones derived from different individuals. As such, they represent hypothetical sequences, not necessarily those utilised to butcher any one animal. The level of consistency across component parts, however, made generalisation possible and productive. I use the results from Cirencester to develop 'Butchery Units' for comparison between parts of the body and between sites and periods. Once the basic framework was established, I incorporate results from the other sites to identify patterns of butchery that could be compared spatially and diachronically. To support the interpretation, I performed replication studies to validate the processes deduced from the archaeological cut marks, using tools that were representative of implements from the two periods.

The demographic, social, and economic developments that followed Roman conquest, many of which subsequently gained momentum and galvanised the formation of medieval society, mark critical points in Britain's past. These faunal assemblages, vast in scale and scope, have been the objects of intense research and offer important evidence for assessing the integral role of animals to these periods. I have chosen to concentrate on the Romano-British and medieval periods in order to compare and contrast between two discrete, well-studied phases. The aim here is not to illuminate trends through time, which has been the basis for numerous other important studies spanning the Iron Age, Roman, Anglo-Saxon, medieval, and post-medieval periods (Maltby 1981, 1985a; Grant 1987; Albarella 1997; see contributions in Serjeantson & Waldron 1989). Rather, I examine the nature of practices in a somewhat more delimited manner, first as a way to provide detailed interpretation and then to propose how the nuances of practice connect to broader socioeconomic activities from these two eras in Britain's history.

The case study focuses on cattle. Within a British archaeological context, cattle served as a multipurpose animal (Dobney et al. 1996: 28; Gidney 2000: 176), providing traction (Bartosiewicz et al. 1997; De Cupere et al. 2000; Murphy et al. 2000: 45), dairy products (McCormick 1992), and manure (Noddle 1989: 40). As deadstock, cattle were variously processed to produce hides (Serjeantson 1989: 129; Murphy et al. 2000: 37), glue (Jones 1985), horn (Yeomans 2008), grease, and of course, meat. The distribution of the animals' flesh reveals not only an economic motivation for raising cattle, but also broader social drivers, with numerous studies illustrating the complexities of division and sharing (Greep 1987: 6; Stokes 2000).

On a more pragmatic level, cattle bones are particularly robust and tend to preserve well, retaining the physical characteristics of cut marks more clearly than species with less dense bone. Their large size necessitates extensive butchery. Research has revealed the varied ways in which cattle were processed. For example, reporting from Greyhound Yard, Maltby observes that, 'as expected, the butchery data show that cattle carcasses were heavily utilised for their meat, skins, marrow and horns' (1993a: 320). However, the main reason for focusing on cattle in this analysis has to do with morphology. Skeletal anatomy is a key driver of butchery (Binford 1978: 48; Lyman 1987; Seetah 2008) and, as such, attempting to decipher the complexity of the butchery process for multiple species would serve only to dilute the resolution.

A BRIEF OVERVIEW OF ROMAN AND MEDIEVAL BRITAIN

The period of Roman occupation represents a complex epoch in Britain's historic past. As a pivotal point that witnessed major transformations in demography, legal structures, and land organization, the period has been well studied and benefits from a wealth of research detailing the everyday lives of the Romano-British people (Collingwood & Richmond 1969; Flower & Rosenbaum 1978; Edwards 1984; James & Millett 2001). There can be no question that the arrival of the Roman army in AD 43, preceded by a significant period of interaction from at least 55 BC (Ireland 2008: 11), had far-reaching implications for Britain (Creighton 2001). Population influx, from a diverse geographic catchment, was both immediate and consistent (see evidence for Africans in Roman Britain, summarised in Eckardt 2014: 63–86). Indeed, Britain's overall population during the Roman period did not again reach the same size until it peaked once again in the Middle Ages. Taking the entire period of 'contact' – not only conquest and occupation – into account, Britain was entangled with the Roman world for some 500 years (Salway 2015: 1).

Perhaps more fundamental are the attendant differences between the type of civilisation in place during the Iron Age and what was to follow. The Romans inaugurated a society that was literate and governed by laws to an extent not observed again until the end of the Middle Ages (Salway 2015: 4). Thus, in some important ways, the Roman and medieval periods share more in common than might at first seem obvious, reinforcing the utility of comparing between periods.

Although historical sources report that Iron Age habitation was 'thick on the ground' (Ireland 2008: 15), there was an increased 'Roman style' urbanisation soon after conquest (Jones 1987). Furthermore, and again despite the fact that Iron Age smiths were proficient, there was an influx of new iron- and steel-making technology (Tylecote 1992: 54). Both of these features had direct and indirect influences on provisioning and consumption, leading, for example, to improvements in cattle (Albarella 1997), intensification of husbandry practices (Maltby 1984; Noddle 1984), and new modes of carcass processing (Maltby 1985a). Roman Britain followed a pattern of meat consumption that differed from that seen in the Mediterranean (King 1999: 189). Cattle were more important than pigs (Dobney 2001: 40), a testament to their versatile uses and significance to the wider economy (Dobney et al. 1996: 30). In Roman Britain, bacon was not the most important meat source (Maltby 1984), and animal husbandry in general mimicked patterns found in other north-western regions of the Empire, as opposed to the Mediterranean south (Dobney 2001: 36). Indeed, King (1999) suggests that rather than 'Romanisation', dietary preference underwent a Gallicisation or Germanisation process, with the army acting as initiator of this transition. However, other aspects of archaeological evidence support a more obvious Roman influence, for example, improvements in cattle that suggest the importation of continental stock (Albarella 1997).

For the Romano-British period, deciphering the intricacies of daily life is a complicated process, particularly where provisioning is concerned, as this depends heavily on cultural as well as economic dimensions. Indeed, any appraisal of meat exploitation per se needs to acknowledge that individuals in 'Roman' Britain represented a diversity of backgrounds with different social norms regarding cuisine and food, as well as a mosaic of local practices and customs. As Hurst succinctly notes with regard to regional identity, 'As for meat-eating, of course there will have been particular breeds of animals and perhaps practices of butchery and consumption with regional characteristics' (Hurst 2005: 303). While there are good textual sources for the continental Empire, similar resources on provisioning are non-existent for Roman Britain. Given the differences between demographic, ecological, cultural, and economic contexts, it is questionable whether Mediterranean sources are useful for the North-West Provinces of the Roman Empire (Grant 1989). These regional disparities across the empire make contributions from animal bone and butchery data all the more important (Grant 1975, 1987, 1989; Maltby 1984).

Following the collapse of Roman hegemony, Britain underwent a protracted period of transition. The Early Middle Ages (600–1066) saw the incursion of the Anglo-Saxons; the High Middle Ages (1066–1272) ushered in a period of Norman rule and radically new attitudes toward labour, land, and property (Carter & Mears 2010: 30–5). With the inauguration of Edward I, the Late Middle Ages (1272–1485) witnessed a reinstatement and consolidation of royal power, which had fragmented along tribal lines with the arrival of the Anglo-Saxons (Bridbury 2008: 81) and become increasingly feudal under the Normans (Carpenter 2004: 84–5). Periods of both civil war and collapse, followed by restoration of law and order, occurred under the rule of successive kings, but during the fourteenth century even greater challenges came from disease, principally the Black Death, and famine (Bridbury 2008: 169–81).

The Late and High Middle Ages are characterised by an overall increase in population and consolidated urbanisation – a burgeoning urban demographic that culminated in an estimated 4 million inhabitants in England by the late thirteenth century (Gillingham & Griffiths 2000: 104). While this led to strife for rural communities who had less land to cultivate and more demands on available land, merchants in the urban environment enjoyed expanding markets (Gillingham & Griffiths 2000: 101). As a consequence of the emerging middle and mercantile classes and extensive economic development (Carlin & Rosenthal 1998: 119), animals took on an increasingly significant role in medieval commerce. Widespread national and international market growth depended on animal products (Thomas 1983; Grant 1988; Astill 1998: 168). More specifically, intensification in the production of light and heavy hides, as well as greater demand for bone and horn, led to a diversification of these subsidiary by-products into distinct industries in their own right (Yeomans 2008). All of these varied enterprises relied on carcass processing in one form or another. By this period, a distinct meat trade was well established with retail outlets, guild associations, and a degree of professionalisation of butchering that included apprenticeship, standardised cutting practices and tools, and documented regulation (Jones 1976; Cowgill et al. 2001: 15).

While Roman metallurgy rarely included steel in utilitarian objects – this more expensive alloy was reserved for weapons – by the medieval period steel technology was commonly included in the fabrication of cutting implements (Cowgill et al. 2001: 10). Concomitantly, the interactions between consumers and producers of such items became more complex. Separate guilds existed for cutlers (Himsworth 1953: 47–60), blade smiths, and so on. As with carcass processing, the labour and craft of blacksmithing also diversified into increasingly more specialised roles.

In concluding this brief overview, while the remainder of this chapter and Chapter 11 discuss 'Roman' and 'medieval' in more general terms, the periods are defined roughly as aligning from the middle to late spectrum of each respective phase. This is based on the broad chronology of the sites under study (see Table 10.1 below).

SITUATING ANIMALS AND FOOD TECHNOLOGY IN ROMAN AND MEDIEVAL BRITAIN

During the Iron Age, butchery relied on extensive use of smaller blades, with few indications of cleavers (Wilson 1978: 119–22; Maltby 1981: 155–7), and tended to favour sites of natural disarticulation, i.e., joints (Wilson 1978: 137; Barber 1999: 115). Furthermore, cut marks on cattle metapodials generally occur on the anterior aspect of the bone (Wilson 1978: 122; Maltby 1985a: 23), suggesting that the skinning process took place with the animal on the ground. Taken together, the butchery has been considered to be 'traditional' (Thawley 1982: 217).

For the Roman period, butchering practices have been well documented (Grant 1971, 1975; Maltby 1979, 1984, 1993a, 1993b, 1996a, 1998a, 1998b; Thawley 1982; Noddle 1984, 2000; Aird 1985; Done 1986; Dobney et al. 1996) and point to a 'Roman' method of processing (Dobney 2001: 39), which followed a recognisably different mode from that seen in Iron Age Britain (Thawley 1982: 217; Maltby 1996b: 22). A consistency in techniques has been noted from sites such as Portchester, Exeter, Cirencester, and Lincoln (Grant 1975; Maltby 1979; Thawley 1982; Maltby 1998c; Dobney et al. 1996). The most distinctive patterns have been observed on urban and military sites. One of the clearest trends distinguishing urban/military techniques from rural is that smaller implements were favoured on the latter (Maltby 1981: 155–204; 1985a: 19–30; 1989: 75–107; 1994: 85–102; 1996a; Thawley 1982: 217). Use of the cleaver can be considered a defining characteristic of urban Romano-British butchery (Coy & Maltby 1984: 84; Maltby 1985a: 20; 1993b). This uniformity has provided strong evidence to support a hypothesis that a special-ised craft was in existence (Grant 1987).

In addition to general trends, more detailed accounts of Roman meat processing have also been constructed from a number of sites. Based on evidence for poleaxing, materials from Portchester (a Roman fort) point to cattle arriving 'on the hoof'. This suggests that the entire process, including slaughter and butchery, took place at the fort, an argument reinforced by the scant evidence for preliminary butchery outside the fort (Grant 1975).

Other characteristics of Roman butchery have pointed to specificity during subsequent processing, for example, meat storage techniques and exploitation of within-bone nutrition. Butchery at the scapula, including trimmed glenoid margins and perforated scapula blades, has been interpreted as evidence of curing the shoulder meat either in brining vats or in smokers (Grant 1975; Dobney et al. 1999). Subsequent knicks to the surface of the scapula blade suggest shaving cuts to remove slices of meat from the shoulder bone (Lau-werier 1988: 156). Evidence for extensive processing of bone for marrow has been observed from Gloucester (Levine 1986), Lincoln (Dobney et al. 1996), and York (O'Connor 1988; Carrott et al. 1995). The butchered assemblages indicate that the meat from most elements of the carcass, and in particular the long bones, was first filleted, and then the bones were systematically chopped. At Lincoln, this pattern occurred on a scale large enough to be considered 'truly commercial' (Dobney et al. 1999). Given the military associ-ations at sites such as Exeter and Portchester (Grant 1975; Maltby 1979), it is probable that the Roman army played an important role in organising the trade in meat (Grant 1987) and disseminating the required knowledge and skills.

Focusing specifically on the tools and techniques of butchery, for the British context the marked transition to cleaver use at Roman military and urban sites should be viewed as a more substantial departure than merely the influx of new forms of processing and cutting tools.

The tools available to the Roman meat processors, as evidenced from the cut marks and directly from the knives (Manning 1976; Grant 1987) fall into four categories: cleavers, large blades, small blades, and saws. Analysis of the extensive array of cutting implements (Manning 1966, 1976, 1985) provides information about the materials present and the technology developed at the time (Cleere 1976; Manning 1979). It is probable that specialist knives were created for the purpose of processing animals. Indeed, Manning suggests that the 'majority of cleavers were probably used for butchering meat' (1985: 120). The key distinction is that the Romano-British cleaver is essentially a dual-purpose tool, designed to slice as well as chop. (The modern British cleaver, by contrast, is employed almost exclusively for chopping.) For the Romano-British period, the distinction between tools used to slice, as opposed to slicing and chopping, could be a useful characteristic for distinguishing between cleavers and general purpose, larger blades.

Finally, we must consider the retail of meat. Research suggests that an organized trade in meat existed in particular regions, for example Cirencester, Walbrook in London, and Exeter (Wacher 1974: 60; Clutton-Brock & Armitage 1977; Maltby 1984: 130; 1985b). Generally, within the Empire, selling occurred from an individual's home, with an opening onto the street forming the main display area (Wacher 1995: 66). Whether this was the same pattern of commercial sale practiced in Britain is hard to verify in light of the available evidence. *Tabernae* (shops and workshops) formed an essential element of both small- and large-scale settlements (Frayn 1995: 107). A number of similarities have been identified between *tabernae* from the continent and Britain. However, one key element, the counter, has yet to be recovered from a Romano-British site (Mac Mahon 2005: 57). Evidence for butchers' shops has been recovered from a number of sites, based largely on the faunal signature, although in some instances this is likely to have been a misinterpretation of normal domestic refuse (Wacher 1995: 405).

In other cases, the evidence is more compelling. For example, large pits with copious quantities of systematically processed animal bone have been recovered from *insula* ii at Cirencester, implying that a butcher's shop or meat market was located within the vicinity (Holbrook 1998: 184–7). Carcass processing requires special consideration with regard to disposal of waste. This might well be a key element in identifying butchers' premises. At Vindolanda, for example, a series of drains would seem to indicate a sluicing system in operation to remove unwanted waste following slaughter (Birley 1977: 40). Shops were not the only retail outlet open to meat traders. The *macellum*, or market, was an important outlet for all manner of produce. At Wroxeter, a large Roman city in Shropshire, an assemblage of bone has been recovered which shares similarities with Cirencester, indicating the presence of a designated meat market (Hammon 2011). Indeed, one of the oldest Roman fora, the *forum boarium*, translates into 'cattle market' (Grant 1978).

From the medieval period, we have considerably more evidence, not only in relation to butchery but also with regard to the human–animal relationship. This has been a boon for helping zooarchaeologists in situating butchery data within the wider societal context. Animals formed part of the visual expression of cultural perceptions and attitudes regarding the natural world and the place of human and beast within this paradigm. Artistic depictions demonstrate scenes of carcass processing following the hunt, such as the butchery of deer in *The Master of the Game* (Baille-Grohman 1904). The act was an important event that culminated the hunt and was evidently a ritualised performance. Archaeological and historical evidence suggests a set pattern of processing was in place, designed to reinforce social inequalities and hierarchy (Sykes 2007).

The rich historical record reinforces the economic significance of butchery, from accounts detailing the organisation of the Butchers' Guild itself (Jones 1976; Rixson 2000: 91–151). At a more specific level, historical records have helped to pinpoint the details of practice. Woolgar (1999: 116), reporting on Goodrich Castle, highlighted that the underlying butchery techniques involved were directed toward preservation and conservation of meat for medium- to long-term provisioning.

Focusing on the analysis of cut marks, numerous researchers have utilised butchery data as a means of assessing broad, multi-period transitions, as well as specific aspects of medieval life (Grant 1987; Maltby 1981). Broad trends have been observed, for example, indicating medial splitting of the carcass (Grant 1987: 56), as noted from Anglo-Scandinavian York (O'Connor 1989b: 155). This tendency has been linked to increased professionalisation of meat processing and the use of specific butchery implements (Grant 1987: 56–7). Other authors have been able to tease out more precise details of daily practice; for example, Wilson (1978: 110–26) used the butchery data to study aspects of trade and provisioning from Oxford. Crabtree (1989: 97), reporting from Anglo-Saxon West Stow, in Suffolk, indicated that chop marks tended to occur near or through joints and that the butchery employed heavy knives and cleavers; the cut marks from York were sufficiently clear to allow for a reconstruction of the overall pattern of dismemberment (O'Connor 1989b: 155). Levitan (1984: 181–8) was also able to reconstruct in-depth butchery and fragmentation diagrams for assemblages from Taunton.

Cut mark data have been used in a variety of ways to study the wider economic significance of cattle by-products, particularly that of the trade in hides and horn. The development of leather working into an organised industry has been closely associated with increased urbanisation (Albarella 2003: 72). Butchers or slaughtermen would have provided the raw material for both the heavy and light leather trades (see Serjeantson 1989: 129–43 and Yeomans 2007: 100–2 for a detailed account of these terms and products). Aside from these economic implications, butchery data have been used to provide insight into culinary practice. For example, the copious quantities of highly fragmented bone from York were suggested to be evidence for large-scale soup production (Bond & O'Connor 1999: 370). Specific patterns of consistent butchery, noted from religious enclaves, have been used to support the notion that religious traditions followed a distinct butchery craft separate from guild practices (Grant 1987: 56).

The medieval period witnessed a great deal of experimentation with type and metallurgic techniques in the manufacture and development of knives. Evidence for this comes from the variety of implements excavated (Cowgill et al. 2001: 78–105), as well as technical factors, such as the way carbon steel and iron was combined to create tools with specific regions of hardness (Cowgill et al. 2001: 10). The techniques used for the creation of metal tools were complex and resulted in implements of considerable durability (McDonnell 1989). Iconographic representations of cleavers illustrate large blades with riveted handles (Fig. 10.1, taken from the *Luttrell Psalter*). A riveted handle increased durability of the tool. Combining steel into the blade resulted in a more resilient and easier to maintain cutting edge.

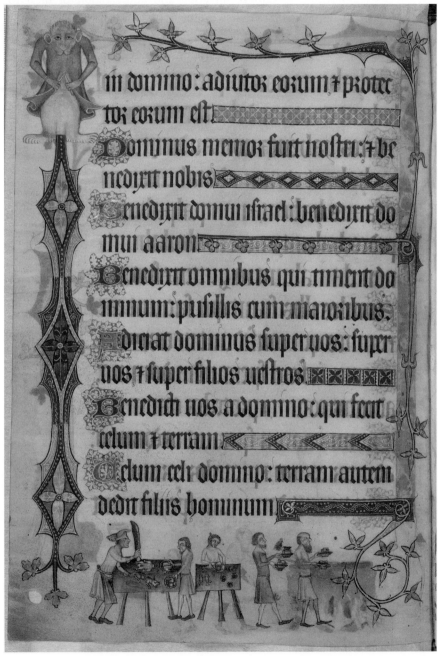

FIG. 10.1 Medieval cleaver depicted in the *Luttrell Psalter*, c. 1325–35. (© The British Library Board, MS. London, B.L., Add. 42130.)

In summary, for medieval cities such as London and York, retail butchery likely followed the methods set out by the respective Butchers' Guilds, the practice passed on during the period of apprenticeship. In turn, these individuals would have been influenced by the culinary requirements of the general populace, the social elite, and during festive occasions (Mead 1931: 78–86; Billington 1990). The guilds established and maintained a standard with regard to cuts of meat and the methods of preparation, which has left a legacy in present-day Britain.

TOWARD A BALANCED REPRESENTATION OF BUTCHERY AND BUTCHERING

The brief overview of archaeological literature on butchery from the Romano-British and medieval periods should signal not only the role and development of the craft but also the complexity of the faunal record from these later-period sites: how do zooarchaeologists unravel the intricate butchery dataset, revealing the nuances of practice for the wider archaeological community?

I now focus on recording of cut marks. The case study has been developed from materials analysed from six sites incorporating a range of settlement types. The inferential process first involved the recording of archaeological butchery marks (Seetah in prep a), followed by experimental replications, helping to test, corroborate, and validate the deductions based on the archaeological data (Seetah in prep c). From these two sets of data, I designated butchery units (subgroups of skeletal elements) to allow for comparisons between sites. Through the archaeological dataset and the experimental studies, I have marshalled both quantitative and qualitative outcomes into a framework that could meet the challenge of providing better inference of cultural, economic, and subsistence features.

An Integrated Approach to Cut Mark Analysis

To summarise from Chapter 9, recording the cut marks involved registering geometric and typological evidence of the mark, indications of the type of tool used to create the mark, and interpretational features denoting butchery activity (namely, distinct characteristics captured within each and every cut mark, although not always recognizable or recordable). From this, three separate approaches to data gathering were generated:

- First, each assemblage provided a large dataset containing evidence of comparative dimensions between sites and periods. This allowed for variability in processing of different carcass parts to be assessed and for butchery patterns to be contrasted.
- Second, by recording the location of the butchery on templates it was possible to generate representative diagrams illustrating groups of marks. The templates

provided a means to visually log clusters of cut marks and monitor the type of butchery pattern that was emerging as the assemblage was recorded. In this way, the templates served to facilitate a reflexive process during data collection. Furthermore, they helped in formulating operational sequences of butchery activity and to schematically outline different approaches to carcass portioning (see Figs. 11.1 and 11.2, Chapter 11).

- Finally, annotations taken during the recording and experimental phase provided evidence of distinctive aspects of butchery procedure. In a small number of cases, it was possible to infer handedness, sharpness of tools, and specialisation in tool manufacture.

Interpreting activity, 'function', enhanced the recording protocol to a state that was representative of actual events and process: the craft as practice. Function was appraised through an assessment of three recorded categories: when the butchery took place relative to disarticulation (i.e., pre/post/during), the carcass position, and the direction of cut (Chapter 9).

Referencing the *chaîne opératoire* approach, functional characteristics were marshalled into stages. These groupings were used to assess sequences and patterns of activity, which were subsequently compared across sites. Three stages were developed. By placing the various functions into stages, it was possible to pool records while maintaining a connection to the original data. The stages are defined in Table 10.2 but effectively pertain to:[1]

- Stage 1: preparatory practice
- Stage 2: flesh-orientated butchery
- Stage 3: bone-orientated butchery.

For each site, there was clear variation in the different categories of function. In order to arrive at a viable means of illustrating the main differences between sites and periods, it was necessary to devise a scheme by which the variability that occurred over the carcass could be represented. This was accomplished by grouping parts of the skeleton into units. The assemblage from Cirencester was large enough for a comparison of butchery activity to be made over the entire carcass, formatted into discrete *chaîne opératoire* sequences (see later). This provided an index of variability across the carcass, which was overlaid with data from the template diagrams. This index provided the basis for the development of the units, rationalised with the benefit of my own expertise.

Combining two approaches for data capture (templates and database recording) mitigated inherent weaknesses in the individual protocols. Templates do not provide a rigorous way to gather all the details of the marks, but they enable rapid assessment of geometric clustering without the need for sophisticated data analysis. In contrast, database systems do not provide the analyst with the visual stimulus that promotes a deeper understanding of the process of butchery but instead promote rapid, detailed data population.

Sampling Strategy and Site Information

Sites were selected based on the following criteria: the size of each assemblage, anticipated quantity of cut marked bone, the level of preservation of material, settlement type, and whether the site had published or in-press reports. With these points in mind, two large, two medium, and two small assemblages were selected (Table 10.1). The sites covered a broad geographic span of Britain, incorporating four counties: Yorkshire (Catterick and Coppergate), Cambridgeshire (Ely), Gloucestershire (Wortley Villa and Cirencester), and Carmarthenshire (Laugharne Castle). In terms of

TABLE 10.1 *Site information, basic counts of bones recorded, and percentages of butchered remains from each case study site (average 27%)*

Site	Site type and period	Σ Records	Σ Butchered	% Butchered	Contextual and other relevant details
ROMAN					
Cirencester	Urban centre with military complex; fourth century	1745	604	34	Thawley 1982; Wacher & McWhirr 1982; Maltby 1990, 1998c; Darvill & Gerrard 1994; MacGregor et al. 1999; Holbrook 1998; Hurst 2005;
Catterick	Town with military outpost and fort; second to third entury	702	173	25	Hilyard 1957; Wacher 1971; Meddens 1990a, 1990b, 2002; Payne 1990; Stallibrass 1997, 2002; Hodgson 2002; Wilson 2002a, 2002b;
Wortley Villa	Villa; second century	353	240	67	Wilson 1986, 1988; Maltby forthcoming
MEDIEVAL					
Coppergate	Urban trading centre; eleventh to twelfth century	921	254	28	O'Connor 1984, 1988, 1989a, 1989b, 1991; Hall 1989, 1999; Bond & O'Connor 1999; Hall & Hunter-Man 2003;
Laugharne Castle	Castle; late twelfth to early thirteenth century	904	187	21	Avent & Read 1977; Avent 1979, 1995; Hambleton & Maltby 2004
Ely	Monastic enclave; fourteenth to fifteenth century	157	44	28	Mortimer et al. 2005; Cessford et al. 2006

settlement type, the sites included urban, i.e., Coppergate, Catterick, and to a certain extent Cirencester; rural, such as Ely, Wortley Villa, and Laugharne Castle; and specialist locations, for example, Wortley Villa and Laugharne Castle.

Variation in sample size was deliberately selected for, in particular as a means of examining the robustness of the methodological approach. However, the limitation of including the smaller samples from Ely and Wortley Villa (n = 157 and 353, respectively) was reflected in the fact that not all skeletal parts were equally represented. The following presents brief background details of the sites studied.

Roman Sites

Cirencester

Cirencester proved to be an ideal test case for an in-depth appraisal of butchery practices as the assemblage was large, with a distinct pattern of cut marks. The Cirencester Excavation Committee managed excavations at Cirencester, 'a model Roman city' (Hurst 2005) with an extensive military complex. The bones studied for this book derive from excavations undertaken in 1980, from Chester Street. Context 5 (P4/L5, 1980/137/5) formed part of the most substantial accumulation from this particular fourth-century layer (site Period 3a), with 3,651 bones recovered, of which 1,968 (54 per cent) were identifiable. The context was composed largely of soil and refuse deposits associated with a late Roman timber building and its subsequent abandonment (Maltby 1998c: 354). Despite some evidence of residual pottery, found in small quantities in the soil, it was evident that this particular deposit was formed from large-scale carcass-processing activities, centred on cattle: c. 2,869 cattle bones were identified from all layers (1–4). Animal gnawing and butchery were the most common modification, and particular attention was paid to the location and types of butchery to occur on cattle bones (Maltby 1998c: 353).

Catterick

Excavations at Catterick, a small town with an associated military outpost, have been carried out since the 1950s. Wacher's excavation of the Catterick Bypass covered some 16 per cent of the Roman fort and town, including a large midden. Copious amounts of animal bone led Wacher to suggest that the site may have been used for large-scale tanning activity (Wacher 1971). The first excavation to systematically recover animal bone was at Dere Street in 1972, subsequently analysed by Payne (1990). Excavations undertaken between 1980 and 1990 recovered large quantities of bone from Bainese (Meddens

1990a, 1990b, 2002) and Thornbrough. The latter site was associated with an Antonine fort and small town (Stallibrass 1997, 2002). Thornbrough Farm (sites 452 and 482) was dated to between the late second to third centuries (site phase 2–5). A total of 4,460 bones (65 kilograms) were recovered with 1,803 identifiable fragments (40 per cent). Site 482 was selected for the present research based on reports of a better level of preservation than site 452 (Stallibrass 1997). Site 482 was composed of a single trench measuring 6 metres by 3 metres in dimension, with a depth of 1.85 metres (Wilson 2002b). In total, 73 separate contexts, from phases 2 to 5, were analysed as part of this study, all of which were associated with the Antonine fort.

Wortley Villa

Paul Cory, a local landowner, accidentally discovered the Roman villa at Wortley in 1981, while digging a fence post-hole. A trench was opened, exposing damaged *pilae* along with a considerable amount of plaster, tegulae (roof tiles), tessarae (mosaic tiles), pottery, and bone. Archaeologists from the University of Keele initially undertook excavations during 1983 and 1984 (Wilson 1986, 1988). This was followed by fourteen further campaigns. The material selected for the present research derived from context 1615 and 1673 and was recovered as part of a three- to four-week training excavation, employing both hand collection and sieving. Overall, preservation of material was considered very good to excellent, with surface modification clearly evident. The two contexts were located in a backfilled cellar deposit, dated to the late fourth century.

Medieval Sites

Coppergate

For the present case, twenty-five separate contexts were studied, with the majority of bone derived from the mid- to late eleventh century, site Period 5Cr. The materials from this site would appear to have been deposited as a result of extensive backfilling. The steep topography of the area apparently encouraged a sustained effort to level large tracts of land through dumping of substrate on low-lying ground. This lasted from the Anglo-Scandinavian to the medieval period (Hall 1999: 307). The assemblages from Period 5Cr were associated with a post-built structure, sealed by a succession of subsequent dumped deposits (Hall 1989: 140). Through dendrochronology, the building was dated to c. 1014–1054 and may have been a warehouse or boat shed. The layer was subsequently sealed by soil, in some parts up to a depth of 2 meters, from the late eleventh or early twelfth century (Hall 1989: 142). The bone material from the medieval deposits was on the whole collected by hand. However, three sites, Coppergate, Tanners

Row, and Fishergate, also underwent wet sieving down to a 2 millimetre mesh. The bone assemblage was substantial, numbering c. 75,000 fragments from all contexts, and noted to be well preserved (Bond & O'Connor 1999: 322–38).

Ely

Excavations at Broad Street offer a valuable opportunity to study both large- and small-scale activity around the waterfront region of Ely. This area was held under the auspices of the monastery during the thirteenth century, which owned some sixty tenements. The site underwent two main seasons of excavation, in 1998 and 2000, prior to development of the area from industrial buildings to an open park linking Broad Street to the River Great Ouse (Cessford et al. 2006). A large faunal assemblage was recovered from the site, with c. 4,364 fragments identified and catalogued. Bone was recovered from periods of occupation spanning the late eighth century to the early nineteenth century. The majority of the excavated material originated from the mid- to late medieval period, accounting for approximately 80 per cent of identifiable fragments, and possibly indicative of the period of greatest activity within the region. Despite the fragmentary nature of the faunal remains, overall the recovered bones were in a good state of preservation. The comminuted state of the sample was attributed to carcass division and butchery practice, rather than post-depositional damage. The bones studied as part of the current research originated from areas A and B and were drawn from seven contexts dated from the fourteenth to fifteenth centuries.

Laugharne Castle

Laugharne Castle is located close to the River Taff, in the south-western corner of Dyfed (formerly Carmarthenshire). The castle was ceded into the guardianship of the Secretary of State for Wales in 1973. Archaeological study began in 1976, and from the outset the site revealed a rich heritage with indications of both prehistoric and Romano-British occupation. How-ever, it is most closely associated with activity from the early medieval to the Tudor period, with post-medieval levels also evident. Excavations of the Inner Ward were undertaken from 1976 to 1978. Some 33,804 frag-ments were scanned from medieval contexts, with the overwhelming majority, 32,921 fragments, derived from the Inner Ward (Hambleton & Maltby 2004). Context 612 was studied for the current research, dated from the latest late twelfth to early thirteenth century. These contexts were associated with a phase of levelling and subsequent rebuilding of the castle and were thus potentially contaminated by residual deposits. However, context 612 (composed of 2,717 fragments, of which 802 (29.5 per cent) were identified) was comprised of an unusually high amount of cattle bone, which was unlikely to be residual.

Developing a Basis for Comparison: Butchery Units

A distinction exists between 'carcass' and 'butchery' units: the former based on natural morphology and body design, the latter on how the animal has been processed. The units outlined in this section are 'butchery units' in a stricter sense and were based on cut mark location and proportions of activity on the various elements. The units were derived from a combination of data: archaeological evidence, the operational sequence, and morphological criteria. Establishing the operational sequence took into account when the specific carcass part (unit) was detached as a separate joint/portion from the rest of the animal. This had to be measured against a set point within the dismemberment process. In this instance, the point of reference was 'further processing': was the portion detached before meat removal, bone breaking, or pot-sizing activity? Finally, the ever-important influence of the animals' morphology was taken into account, incorporating the ease or difficulty with which different parts of the carcass were disarticulated. Amalgamating data permitted comparisons to be made between sites and periods, testing whether variations were evident in the types of tools used for different aspects of the butchery process. However, there was an increased masking effect that likely resulted in a loss of variation between individual body parts. The descriptions for units, function, and stages are summarised in Table 10.2. Applying the same units across all the study assemblages was justified on the basis that the sites shared a number of similarities. This included being

TABLE 10.2 *Summary of units, functions, and stages*

Parameter	Description
Unit 1	Head and cervical vertebrae
Unit 2	Thoracic vertebrae and ribs
Unit 3	Scapula, humerus, radius, and ulna
Unit 4	Lumber, sacral and caudal vertebra, and pelvis
Unit 5	Femur and tibia
Unit 6	Lower fore-limb (carpals and metacarpals)
Unit 7	Lower hind limb (tarsals and metatarsals)
Function 1	Disarticulation
Function 2	Meat removal
Function 3	Filleting
Function 4	Prep mark
Function 5	Bone breaking
Function 6	Skinning
Function 7	Jointing
Stage 1	Evisceration, skinning, or gross disarticulation (functions 1, 4, and 6)
Stage 2	Jointing and meat removal (functions 2 and 7)
Stage 3	Pot-sizing and/or final filleting (functions 3 and 5)

comparable, as groups, on a regional and chronological basis and in terms of the types of technology used during butchery.

DISCRETE *CHAÎNE OPÉRATOIRE* PATTERNING: BUTCHERY AT CIRENCESTER

Cirencester served as a 'gold standard' case study, with 1,745 individual bones recorded (604 with cut marks, 34 per cent), as per Table 10.3. The assemblage was suitably large, allowing for outlines of specific activity patterns to be developed based on morphological divisions of the carcass. The frequency data, combined and presented as stages (illustrating a range of activity by highlighting variations in how different carcass parts were treated), are discussed in the next section but are presented here, as they were used to define the sequence.

Axial Skeleton

The skull was represented only by elements from the mandible. The absence of cranial bones resulted in an under-representation of specific marks indicative of skinning and head removal. Butchery of the mandible centred on two main activities: disarticulation of the jawbone from the head, evidenced through point insertion and knick marks around the mandibular condyle, followed by extensive bone breaking of the mandible using the cleaver. The conspicuous absence of the cranium, coupled with a representative sample of the mandible, suggests that these two components of the skull were treated in different ways. There was clear consistency in the size and portions into which the mandible was butchered, indicating a set pattern for cut placement and specific requirements from the portion. Part of the cranium, specifically the horns, may have had value for working. Other resources, such as the brain, may also have been differentially processed.

The cervical vertebrae, more so than the other vertebrae, were characterised by a predominance of fine blade marks, indicating gross dismemberment. Bone breaking activity appears to have occurred after gross dismemberment and meat removal, as evidenced by chop marks from a number of angles, along both transverse and axial planes. This indicates repositioning during butchery, which tends to occur once the portion has been reduced in size and can be easily manipulated. At least one example indicates that the head was removed with the animal on the ground. A blade insertion behind the ventral aspect of the cranial articular surface of an axis suggests that the head was extended to expose the neck, and a blade inserted to decapitate the head. In this instance the atlas was grouped with the rest of the head. It is likely that this region, and possibly the rest of the skull, had already been skinned; otherwise, we would observe evidence of point insertion marks, rather than blade insertions, as the

TABLE 10.3 *Frequency of butchery activity from Cirencester per element*

Element → Function	Hd	Cv	Tv	Rib	Sca	Hum	Rd	Ul	Lv	S/Cv	Pel	Fem	Tib	Lfl	Lhl
1: Disarticulation	40	34	27	10	12	20	7	8	1	1	4	5	2	1	106
2: Meat removal	4	4	15	36	85	10	15	5	2	1	30	61	76		22
3: Filleting			6	55	6						2	23	1		14
4: Prep mark									1	1	2	16	2		
5: Bone breaking	22	13	41	35	1	10	12	10		2	23	20	12	1	28
6: Skinning														1	9
7: Jointing															
Σ FUNCTION	**66**	**51**	**89**	**136**	**104**	**40**	**34**	**23**	**4**	**5**	**61**	**125**	**93**	**3**	**179**
Stage 1 (%)	60	67	37	7	11	50	21	35	50	40	10	17	4	75	64
Stage 2 (%)	6	8	17	26	82	25	44	22	50	20	49	49	82	0	12
Stage 3 (%)	33	25	46	66	7	25	35	43	0	40	41	34	14	25	42
Σ bones	**220**	**39**	**99**	**505**	**146**	**62**	**70**	**76**	**11**	**14**	**52**	**116**	**62**	**61**	**222**
% Butchered	**30**	**130**	**90**	**27**	**71**	**66**	**48**	**30**	**36**	**35**	**117**	**107**	**150**	**5**	**81**

For each element, the sum of the different types of butchery activity was recorded. The exception to this was lower front limb (Lfl, carpals and metacarpals), lower hind limb (Lhl, tarsals and metatarsals), and sacral and caudal vertebrae (S/Cv). Other abbreviations: Hd, head; Cv, cervical vertebrae; Tv, thoracic vertebrae; Sca, scapula; Hum, humerus; Rd, radius; Ul, ulna; Lv, lumbar vertebrae; Pel, pelvis; Fem, femur; Tib, tibia.

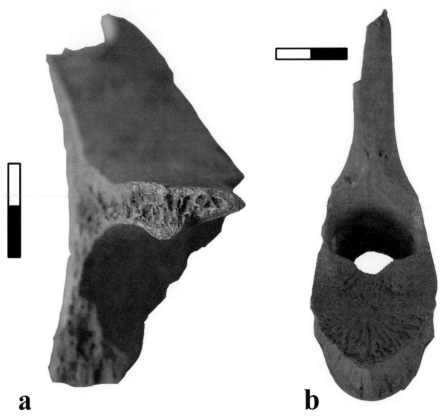

FIG. 10.2 Top-down blade insertion for thoracic spine sectioning (a); axial splitting of thoracic vertebrae on either side of vertebra (b).

point of a knife would be used to puncture the skin and commence the decapitation process.

The thoracic vertebrae had undergone a distinctive and repetitive pattern of breakage that followed gross dismemberment and meat removal. This section of the spine was removed as a block and subsequently processed with a combination of blade and cleaver butchery. To accommodate the long thoracic spines, a specific technique was employed to portion this part of the carcass. The thoracic spines are angled; a direct chop from above will encounter at least two separate vertebral spines through its arc of descent, which has the effect of cushioning the blow, preventing a clean chop. To avoid this, a blade can first be inserted at the top of one thoracic spine and pushed down. This exposes the vertebral joint and provides a clearer target for cut placement of the subsequent chop (see Fig. 10.2a). This technique can be used along the length of the spine and is effectively a 'prep mark' for jointing the vertebral column. It can be used for individual vertebra or sections of vertebrae.

Evidence that the spine had been removed in a block comes from the fact that it had been split on either side of the main body of the vertebra, removing the lateral spines and articulations with the rib heads (see Fig. 10.2b). This must have occurred after the transverse splitting described above, otherwise there would have been no need to split the spine on either side of the centrum as the vertebrae would already be in smaller components. Furthermore, we would not observe the same regularity as noted from this assemblage. With smaller portions of bone and meat there would have been a greater degree of repositioning and less systematic chopping. The vertebrae must have still been articulated, imposing a morphological constraint to the butchery, resulting in consistent chopping on either side of the spine.

The lumbar vertebrae showed a pattern of butchery unique to this section of the spine. There was no direct evidence of bone breaking; however, dismemberment and meat removal were equally well represented. The point to stress here is that while chop marks per se were not identified, evidence for chopping was present in the form of vertebral body (cancellous bone) with straight and clean fracture lines. Dismemberment and meat removal marks illustrate the removal of flesh from both dorsal and ventral aspects of the lumbar vertebrae. Indeed, point insertion marks on the ventral surface indicated removal of the kidneys and the soft 'long fillet'.

The sacral and caudal region of the spine was represented by relatively few bones (fourteen), of which no recordable butchery was noted. Butchery was deduced indirectly from the fracture lines on these bones. Marks on the caudal vertebrae illustrated that a similar technique for sectioning the thoracic spine was used for tail removal: inserting a blade between the vertebral bodies and detaching the tail from the rest of the spine. This method requires detailed knowledge of the animal's anatomy and precise cut placement. The region into which the blade needs to be inserted is approximately 2 millimetres wide, on either side of which are the vertebrae. This action can be performed with a cleaver but requires an even greater level of accuracy as a chopping cut needs to be applied (invariably from a greater distance) to the same point between the joints of the caudal vertebrae. Furthermore, this can only be performed on a chopping block. Thus, as only blade marks were noted from these bones it is apparent that the butchers at Cirencester were removing the tail relatively early within the dismemberment process, perhaps during skinning. This is further supported by the fact that at least one of the caudal vertebrae with this type of cut mark was the first or second after the sacral vertebrae, indicating that the tail may have been removed as an entire unit.

Alongside the vertebrae, the ribs provided evidence for the specific techniques used for gross disarticulation. In particular, a range of marks – from 'shaving' to near total removal of the rib head – indicated that the flanks had been removed from the vertebrae with the carcass suspended, or the thorax placed on a chopping block, cranial end down. As discussed above, by chopping on either side of the spinal column, the lateral spines of the vertebrae, along with part of the rib heads, were detached. Aside from this, the ribs provided some intriguing indications of culinary practice. The bones of the rib cage were processed consistently into pieces approximately 10 centimetres long, often with chops to either side of the bone, a practice descriptively termed 'pot-sizing' (Rixson 1988). A more recent correlate has been noted from the nineteenth- to early twentieth-century site of Fort Johns in Sussex County, New Jersey (Crabtree & Campana 2008: 324).

Upper Fore-Limb

Meat removal activity predominated within this region of the carcass. A considerable amount of flesh encases the awkwardly shaped scapula bone, and deboning is complex and time consuming. This was reflected in greater variability of tool use. Small and larger blades were used to disarticulate the scapula and humerus and to remove meat. The cleaver was employed to chop the bone and, interestingly, for removing the scapula spine; this may have occurred while meat was still on the bone. Although this particular activity is not unique to the Romano-British period (similar if not identical marks have been observed from Coppergate), the consistency with which it is noted on Romano-British sites, in particular military enclaves, has resulted in this mark being considered typical of Romano-British butchery (Dobney et al. 1996: 26–7). Why this particular technique was favoured is a complicated question to answer (see Seetah 2002). It would appear that speed and a desire to maintain the overall shape of the shoulder, to facilitate suspending the joint, under-pinned this activity. This may have improved flavour or facilitated subsequent cutting, i.e., filleting fine slices from the cured joint.

Butchery of the humerus, radius, and ulna (Fig. 10.3) illustrated the typical signature of Romano-British military butchery, where the cleaver was used for both disarticulation and subsequent bone breaking. Using the cleaver in this manner illustrates the desire to expedite the butchery process. In all three cases the disarticulation of these elements from each other could have been carried out with a less specialised knife. However, this would have been more time consuming. The Romano-British butchers opted for the more labour-intensive alternative, using powerful chopping blows to disarticulate the fore-limb into its constituent parts. These chops required a relatively high degree of force to deal with the thick cortical surface; the distal radius for example is a particularly dense area of bone. The force needed was generated through the effective use of momentum; the blows were delivered with precise cut placement.

To take the example of the distal radius, oblique chops were delivered above the joint, thus taking advantage of two specific aspects of bone archi-tecture at this region (Fig. 10.3a). First, the region of the metaphysis (region of the bone forming a margin between the articular head and the main cortical

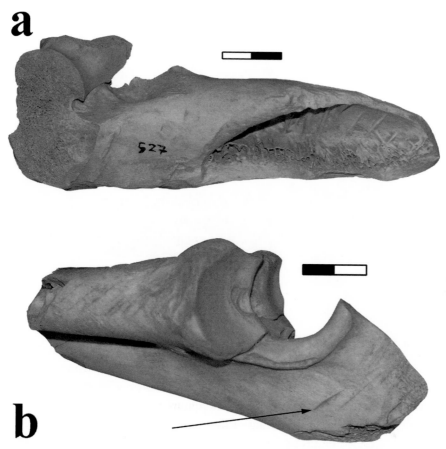

FIG. 10.3 Use of the cleaver to disarticulate the fore-limb: (a) humerus, (b) use of prep marks (arrowed) to locate the articulation of the ulna and humerus prior to chopping.

bone) above the joint encases cancellous bone and is less dense than the surrounding area. Second, cortical bone is more resilient in the transverse plane than the axial. It would not be possible to cut the bone along its length while it was encased in flesh. By exposing an area of bone and chopping into it at an angle, the butcher can take advantage of the bone's inherent weakness. The same technique was used for disarticulation at the distal humerus. Prep marks within this region, particularly at the ulna, were evident, indicating the need to first locate, and then chop the joint (Fig. 10.3b).

Pelvis and Upper Hind Limbs

Butchery around the pelvis bones (the innominate) was dominated by meat removal and bone-breaking activity, which followed detachment from the femur, as indicated by disarticulation marks. This specific pattern reflects the need to pare flesh from an awkwardly shaped bone, followed by subsequent fracturing in order to reduce the overall size of the pelvis bones (the

FIG. 10.4 Typical fractured femoral head.

innominates) to fit into a cooking vessel. The disarticulation technique used at the femur provided a way to better understand the principles, and rationale, governing the butchery of this site. A chopping blow was utilised to disarticulate the legs *prior* to axial splitting of the spine. Disarticulating the legs in the manner described results in a fractured femoral head, as illustrated in Fig. 10.4. This provided a clear indication of the operational sequence of butchery that took place at Cirencester. Marrow extraction from the femur and other long bones was also important, as evidenced from large numbers of axially fractured shafts, coupled with consistent fracturing of the distal epiphysis.

Butchery at the tibia was predominated by meat removal marks, made with a variety of tools and techniques. For example, point insertions using the tip of a sharp blade were used to detach meat by cutting into the attachments connecting flesh to bone (a technique termed 'muscle boning') (Fig. 10.5a). This was coupled with blade insertion marks, whereby a larger portion of the knife edge is drawn down the length of the bone to pare away flesh. Invariably, this is performed while the joint is being held, with one end resting on the cutting block. This has been demonstrated archaeologically; the direction of the cut travelled along an inferior-superior trajectory, indicating that the proximal end of the bone was resting on a butcher's block, and the distal end was held uppermost (Fig. 10.5b). Subsequent to meat removal the bone was fractured along the axial plane for marrow extraction. We can be confident that this was the correct sequence of events as otherwise there would have been no need for muscle boning; the meat would have been removed after cooking without leaving these types of marks.

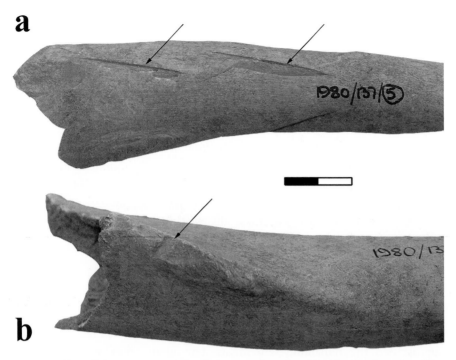

FIG. 10.5 Two distinct meat removal techniques used on the tibia: (a) using the point of the knife, (b) using a portion of the blade.

Limb Extremities

The cleaver was used to disarticulate the lower extremities, particularly for the hind limb, which in itself is intriguing, as this is one instance where it would have been almost as rapid to use a general-purpose knife to remove the lower limbs. What appears to have prompted the use of the chopping action of the cleaver, for the hind limb at least, is that the lower limb was removed relatively high, at the uppermost aspect of the hock joint, below the tibia. This activity pattern was noted on the calcanei and astragali (the two main bones of the ankle joint), which were fractured transversely, with the cut travelling from the lateral to medial facet. This contrasts with modern techniques where a skinning or boning knife is used to remove the lower limbs beneath the carpals/tarsals (the small bones of the wrist and ankle joint), slightly above the metapodials.

In summarising this section, there are indications of each 'stage' of processing on almost all carcass elements, with the exceptions of the limb extremities. However, there are distinct variations in the degree to which the different categories of butchery activity occur on individual portions of the carcass. Certain practices were highly idiosyncratic and thus revealing. For example, use of the cleaver to remove the scapula spine appears to relate specifically to curing practice, as discussed in greater detail in Chapter 11. Storage and preservation can dictate the types of cutting practices employed. A minimal amount of cutting is performed prior to curing in order to reduce the likelihood of bacterial infiltration. Deboning the scapula would result in more exposed flesh. By simply trimming the scapula spine, the bulk of the meat is unaffected and can be cured with minimal risk of spoilage. The copious quantity of rib portions, numbering some 505 individual fragments and invariably portioned to a specific size, suggests larger-scale resource extraction. In general, bone breakage for marrow extraction was prevalent, reinforcing the notion that exploitation of a broad range of resources, including fat and grease, were key features of butchery at Cirencester. A final point to reinforce is that the butchery noted from the spine and hind limbs suggests that the overall pattern revolved around piecemeal disarticulation of main carcass parts, followed by further processing performed on a block.

BUTCHERY REPRESENTATION USING FREQUENCY OF OCCURRENCE

Building on the detailed assessment of marks from Cirencester, I applied the approach to the remaining settlements. For all the sites, each mark was recorded as a single count, including incidences of repetition, at any single location. This provided an accurate measure of frequency, as opposed to only the occurrence of a particular type of butchery. In this way butchery patterning could be pinpointed by looking at variability over the carcass (represented through 'frequency of activity'). Once the frequencies of different types of butchery activity were noted, they were placed into their respective categories according to function. Each stage was then defined. Comparisons between sites were made on the basis of variation in activity across the carcass, per butchery unit. The same butchery units were employed for comparison of implement use. This latter dataset provided an additional means of assessing differences in the techniques used during dismemberment and whether this variation was evident in the implements and their proportions of use, and incorporated those marks that could not be assigned to a function. In this way, the approach makes more effective use of the global dataset. In combination, this led to a clearer appraisal of nuance within the butchery process. It was occasionally possible to infer the use of the Romano-British cleaver, for example, when the 'prep-chop' method was observed. However, although recorded separately, in the following summary indications of Romano-British cleavers are discussed under the 'cleaver' category. Tables 10.4–10.7 show the occurrence of butchery activity and implement use as they were represented for each unit.

Generalising from the tables, variations in butchery patterning are evident across the carcass and between sites, based on frequency of cut marks. Each

TABLE 10.4 *Butchery activity per unit, Romano-British sites*

Units → Function	1	2	3	4	5	6	7
CIRENCESTER							
1: Disarticulation	74	37	47	6	7	1	106
2: Meat removal	8	51	115	33	137		22
3: Filleting		55	6	2	24		14
4: Prep mark		6		4	18		
5: Bone breaking	35	76	33	25	32	1	28
6: Skinning						1	9
7: Jointing							
TOTAL	117	276	201	70	220	3	179
Stage 1 (%)	63	15	23	14	12	66	6
Stage 2 (%)	7	18	57	47	63	0	13
Stage 3 (%)	30	66	20	39	25	33	23
Σ Bones	**259**	**604**	**308**	**77**	**178**	**61**	**222**
CATTERICK							
1: Disarticulation	29	8	18	5		2	16
2: Meat removal	10	42	81	1	9		
3: Filleting							
4: Prep mark							
5: Bone breaking	11	34	4	16	8		1
6: Skinning	3						18
7: Jointing		6					
TOTAL	60	90	103	22	17	2	35
Stage 1 (%)	53 ± 12.6	10 ± 6.2	17 ± 7.3	23 ± 17.6	0	100	97 ± 5.7
Stage 2 (%)	17 ± 9.5	53 ± 10.3	79 ± 7.9	5 ± 9.1	53 ± 23.7	0	0
Stage 3 (%)	30 ± 11.6	37 ± 9.9	4 ± 3.8	73 ± 18.6	47 ± 23.7	0	3 ± 5.7
Σ Bones	**167**	**166**	**142**	**40**	**34**	**27**	**85**
WORTLEY VILLA							
1: Disarticulation	4		27		33	4	11
2: Meat removal	11	37	67		91		
3: Filleting							
4: Prep mark							
5: Bone breaking	6	53	29	5	65	1	5
6: Skinning			3				20
7: Jointing							
TOTAL	21	60	126	5	234	5	36
Stage 1 (%)	19 ± 16.8	0	24 ± 7.5	0	15 ± 4.6	80 ± 35	86 ± 11.3
Stage 2 (%)	52 ± 21.4	62 ± 12.3	53 ± 8.7	0	58 ± 6.3	0	0
Stage 3 (%)	29 ± 19.4	38 ± 12.3	23 ± 7.3	100	27 ± 5.7	20 ± 35	14 ± 11.3
Σ Bones	**15**	**68**	**84**	**3**	**96**	**3**	**20**

TABLE 10.5 *Implement use per unit, Romano-British sites*

Units ⟶ / Implement	1	2	3	4	5	6	7
CIRENCESTER							
Cleaver (%)	48	33	25	15	29	50	17
Large blade (%)	52	27	36	36	57		45
Fine blade (%)		39	34	15	6	50	29
Romano-British cleaver (%)		1	5	33	7		9
Σ PER UNIT (%)	10	27	24	6	21	<1	16
Σ cut marks	**122**	**320**	**276**	**66**	**246**	**2**	**184**
CATTERICK							
Cleaver (%)	33	52	28	76	53		6
Large blade (%)	54	48	56	24	47	50	17
Fine blade (%)	13		9			50	77
Romano-British cleaver (%)			7				
Σ PER UNIT (%)	20	27	29	7	5	<1	13
Σ cut marks	**69**	**94**	**98**	**25**	**19**	**2**	**35**
WORTLEY VILLA							
Cleaver (%)	25	20	24	23	35	80	19
Large blade (%)	49	38	44	57	52		70
Fine blade (%)	26	23	17		4		8
Romano-British cleaver (%)		19	15	20	9	20	2
Σ PER UNIT (%)	5	21	25	11	28	<1	10
Σ cut marks	**35**	**149**	**178**	**79**	**202**	**5**	**74**

major division of the carcass – the axial parts, fore-limb, and hind limb – is treated differently and this too varies across sites and between periods. Implement use is also variable, with a preference for larger blades, including cleavers. While one might expect this based on the fact that these tools leave more prominent and easily recognised modifications, the variability also correlates with the fragmentation pattern, and the frequency of specific tasks, such as filleting.

Thus, in summary, Romano-British butchers appear to have favoured large blades and cleavers. However, all three sites showed markedly different percentages of implement use. From Cirencester, the greater proportion of final filleting marks (noted on the lateral surface of the ribs) were reflected in a high percentage of fine blade marks, while pot-sizing activity was observed in the predominance of cleaver use. Meat processing at Catterick depended on large blades and cleavers exclusively, complementing the butchery data that outlined

TABLE 10.6 *Butchery activity per unit, medieval sites*

Units ⟶	1	2	3	4	5	6	7
Function							
COPPERGATE							
1: Disarticulation	16	25	26	14	19	4	2
2: Meat removal	2	92	13	32	6		
3: Filleting							
4: Prep mark				1			
5: Bone breaking	27	83	10	37	4		
6: Skinning	6					5	5
7: Jointing							
TOTAL	51	200	49	84	29	9	7
Stage 1 (%)	43 ± 13.6	13 ± 4.7	53 ± 13.9	18 ± 8.2	66 ± 17.2	100	100
Stage 2 (%)	4 ± 5.4	46 ± 6.9	27 ± 12.4	38 ± 10.4	20 ± 14.6	0	0
Stage 3 (%)	53 ± 13.7	41 ± 6.8	20 ± 11.2	44 ± 10.6	14 ± 12.6	0	0
Σ **Bones**	**204**	**372**	**57**	**86**	**50**	**18**	**67**
LAUGHARNE							
1: Disarticulation	16	10	16	11	17	19	20
2: Meat removal		64	13	37	10		
3: Filleting							
4: Prep mark							
5: Bone breaking	10	23	5	31	2		3
6: Skinning							56
7: Jointing							
TOTAL	26	97	33	79	29	19	97
Stage 1 (%)	62 ± 18.7	10 ± 5.9	49 ± 17.1	14 ± 7.7	59 ± 17.9	100	76 ± 8.5
Stage 2 (%)	0	66 ± 9.4	39 ± 16	47 ± 11	34 ± 17.2	0	0
Stage 3 (%)	38 ± 18.7	24 ± 8.5	12 ± 11.1	39 ± 10.8	7 ± 9.3	0	3 ± 8.5
Σ **Bones**	**105**	**243**	**94**	**91**	**74**	**38**	**116**
ELY							
1: Disarticulation	2			2	3	2	11
2: Meat removal	1		15	8			
3: Filleting							
4: Prep mark							
5: Bone breaking	16	6	3	10	6	3	
6: Skinning							
7: Jointing		1					
TOTAL	19	7	18	22	9	5	11
Stage 1 (%)	11 ± 14.1			10 ± 12.5	33 ± 30.7	40 ± 32	100
Stage 2 (%)	5 ± 9.8	14 ± 25.7	83 ±17.4	40 ± 20.5	0	0	0
Stage 3 (%)	84 ± 16.5	86 ± 25.7	17 ±17.4	50 ± 20.9	66 ± 30.7	60 ± 32	0
Σ **Bones**	**69**	**16**	**12**	**16**	**14**	**7**	**16**

TABLE 10.7 *Implement use per unit, medieval sites*

Units →	1	2	3	4	5	6	7
Implement							
COPPERGATE							
Cleaver (%)	58	48	27	51	20		14
Large blade (%)	25	42	55	33	80	33	29
Fine blade (%)	17	10	18	16		66	57
TOTAL PER UNIT (%)	13	45	11	20	7	2	2
Σ **cut marks**	**59**	**199**	**49**	**87**	**29**	**9**	**7**
LAUGHARNE							
Cleaver (%)	64	29	32	53	15		4
Large blade (%)	18	54	26	28	82		23
Fine blade (%)	18	17	41	19	3	100	73
TOTAL PER UNIT (%)	6	26	9	21	9	6	22
Σ **cut marks**	**22**	**95**	**34**	**78**	**34**	**21**	**80**
ELY							
Cleaver (%)	57	100	17	50	91	60	
Large blade (%)	26		83		9		73
Fine blade (%)	17			50		40	27
TOTAL PER UNIT (%)	24	8	19	21	11	5	11
Σ **cut marks**	**23**	**7**	**18**	**20**	**11**	**5**	**11**

near-equal proportions of meat removal (large blade) and pot-sizing (cleaver) activity. At Wortley, the activity pattern favoured meat removal, as well as indications of Stage 2 activity. These patterns are reflected in the implements data, demonstrating a greater degree of variability across the tool categories.

In contrast to the Romano-British assemblages, the results from the medieval assemblages fell within a smaller range for overall percentage of butchery activity per site. For the head, for example, the butchery activity at Laugharne followed the anticipated pattern: the focus was on skinning and gross disarticulation of this high bone/low meat part of the carcass, with Stage 1 activity dominating. However, the butchery at Coppergate indicated a higher level of Stage 3 activity, reflecting proportionally more fracturing and filleting marks. This, as in the case with Cirencester, would be in keeping with a more intensive mode of within-bone nutrient exploitation. The results describing implement use generally indicate that the medieval butchers favoured larger blades and cleavers. However, there is a greater degree of variability, with more indications of fine blade usage, when compared with the findings from the Romano-British sites. The broader implications of these results are developed and discussed in Chapter 11.

VALIDATING 'PROCESS' THROUGH EXPERIMENTATION

The replication studies defined the sequence of butchery, a *chaîne opératoire* in the more traditional sense, as deduced from the archaeological marks (see Seetah in prep c). I undertook two sets of experimental replications: the first to produce knives that would then be used in the second set of replications, to assess the butchery process. Thus, in reference to the guiding principles identified in Chapter 3, an object in the form of a cleaver was produced to deconstruct another 'object', the carcass, to produce a third commodity, meat. Only cleavers were replicated, one each from the Romano-British and medieval periods. It is relatively difficult to identify specific butchery knives from the archaeological record. However, cleavers show many features that are ideally suited for processing carcasses and they appear in numerous iconographic depictions showing butchery of a range of fauna.

Professional blacksmiths, who specialise in archaeological techniques of forging, reproduced the two cleavers. Their methods included the use of bellows as opposed to an electrically powered gas-fired forge, using traditional hammer and anvil techniques rather than a powered hammer or press, and actually forging the overall shape of the tool while in the furnace rather than simply grinding the implement from a blank piece of pre-fabricated iron. The Romano-British cleaver was commissioned from Hector Cole and was manufactured at his forge, Couzens Farm Studios, in Little Somerville, Wiltshire. Steel was not used in the production of the Romano-British cleaver. The tang was left as an open flange and then wrapped around a handle fashioned from oak. The handle was fixed in place with rivets. The blade had a strong curve and was pointed, thus replicating the typical shape of cleavers from this period.

Creating the medieval cleaver formed part of a project carried out at the Lejre Centre, Denmark, in August 2003. The project took advantage of the Lejre Centre's forging facilities as well as the blacksmithing skills of Aron Hvid to recreate the desired tool. The basic design of the tool was taken from a combination of archaeometallurgic finds and iconographic representations, principally a depiction of a cook/butcher using a cleaver with a riveted handle, taken from the *Luttrell Psalter* (Fig. 10.1).

Furnished with the correct cutting tools, and based on evidence from the recorded assemblages, a suite of cut marks was identified that could be used to deduce the butchery process. However, as a consequence of the UK Department for Environment, Food and Rural Affairs (Defra) regulations following the bovine spongiform encephalopathy outbreak in Britain, I was unable to obtain an intact cow carcass that had not been split for removal of the spine and other cerebral material. The best solution was therefore to use a red deer (*Cervus elaphus*), which I procured from the Ross-shire Forestry Commission, Scotland. The animal was a young stag, approximately one and a half years old, weighing approximately 40 kilograms dressed (eviscerated, and with the feet and head removed), but with the skin (which accounted for 8 kilograms). I estimate the animal would have weighed at least 65–70 kilograms prior to removal of the head, feet, and internal organs. It had been shot through the chest/abdomen by James McCloud, a Forestry Commission Ranger, three days prior to collection. As it was possible to procure only one carcass, an operational sequence for both periods had to be carried out on the same carcass. This was only a problem for initial splitting and disarticulation; for the majority of the remaining cuts, this situation did not pose an issue as half a carcass was reserved for each period.

The protocol for the replications was derived from earlier studies (Sadek-Kooros 1972; Jones 1980; Binford 1981; Stanford et al. 1981; van Wijngaarden-Bakker 1990) but based principally on personal experience. I made no attempt to replicate specific marks; rather, I was interested in qualifying the sequence of butchery deduced from the archaeological record. In many cases, the archaeo-logical marks were identified as distinct, but not unique, to one or other of the two periods under investigation. The majority of the marks for the Romano-British period were taken from Cirencester, while the corpus of period-specific butchery marks for the medieval period derived from Coppergate. However, in both cases, material from all sites was utilised.

Summarising from the replications, in general terms, disarticulating the carcass into large portions required a relatively small amount of butchery, whereas to process smaller joints for cuisine requires a proportionally higher number of cuts. Only one technique could be employed for the spine, used to establish the operational sequence for the first carcass separation. This division formed the precursor to subsequent butchery. In this case, it was apparent that axial splitting and femoral disarticulation are mutually exclusive, serving as an important distinction between the periods, and discussed in greater detail in Chapter 11. Butchery of the pelvis offers an important example of one of the more important drivers: cuisine. Despite being composed mainly of cortical bone, the pelvis was treated in the same way as the vertebrae, composed predominantly of cancellous bone. It was fragmented into small pieces and probably processed for fats.

INTERPRETING QUANTITATIVE AND QUALITATIVE EVIDENCE

Romano-British Patterns of Butchery

A combination of cleaver and knife butchery took place on the Romano-British sites to detach the main elements of the fore-limb into their constituent parts. From the archaeology it was clear that the main technique for detaching the humerus from the scapula employed a blade, as this is a relatively mobile joint and

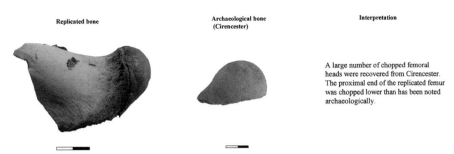

Replicated bone   **Archaeological bone (Cirencester)**   **Interpretation**

A large number of chopped femoral heads were recovered from Cirencester. The proximal end of the replicated femur was chopped lower than has been noted archaeologically.

FIG. 10.6 Hind limb disarticulation.

easy to disarticulate. However, this leaves relatively few cut marks, making it hard to establish a specific disarticulation technique. The techniques for dismemberment of the other elements from this region of the carcass were easier to ascertain. Use of prep marks followed by cleaver chopping predominated, employed to detach the lower from the upper fore-limb slightly above the point where the ulna articulated with the olecranon fossa of the humerus. The distal radius was chopped obliquely, above the articular facet, thus taking advantage of the inherent weakness of the thin cortical bone in this region (similarly to the example shown for the humerus, Fig. 10.3a). Of particular interest is that this strategy of butchery results in fracturing of the joint ends, which would then have been relatively easy to fragment further for grease and fat exploitation. Whether this was a deliberate activity pattern to facilitate further processing or a by-product of changing tool technology is unclear, but it could potentially serve as an index to qualify the efficiency of the overall butchery process.

The same general principle of prep mark followed by cleaver chopping was also employed for the hind limb. The sequence of butchery highlighted by Fig. 10.6 illustrates that the butchery replications I performed do not correspond with the practices evidenced from the Romano-British assemblages. The analogous archaeological femoral head shows that the femur was chopped slightly below the neck of the femoral head. As half of the carcass was assigned to each period, I split the carcass axially along the midline. This resulted in the femur being detached with a portion of the pelvis; consequently, there was no need for femoral disarticulation by chopping into the femoral head. The approach to femoral disarticulation noted from the archaeological examples is distinctive and was performed following a specific sequence. Although the same technique was not replicated, splitting the carcass axially resulted in disarticulation of the femur, offering convincing evidence that the approach replicated here was not the one used on the Romano-British sites that were studied. This particular technique offers the most conclusive evidence for 'piecemeal' carcass disarticulation. The cleaver was used extensively for the remaining carcass butchery of the hind limb, including disarticulation of the hock joint and for long bone and joint-end fracturing for marrow and grease extraction.

Medieval Patterns of Butchery

The techniques used for sectioning and further processing of the vertebrae were not as clearly apparent from the medieval as they had been from the Romano-British assemblages. The cleaver was certainly in use; however, it would appear that sectioning by cutting into the intervertebral joints occurred with a knife, rather than with the cleaver. In contrast to the 'prep-chop' activity of the Romano-British sites, the medieval butchers appear to have relied on a selection of blades as the principal disarticulation tool. The cleaver was used extensively for bone breaking activity, but generally speaking, there was greater emphasis on blade butchery in order to produce portions of a size suitable for cuisine. This was particularly prevalent on the fore-limb. Procedures that utilised the cleaver on the Romano-British sites were carried out with a blade on the medieval sites. Meat removal and jointing processes were also performed with various blades.

The same preference for blade cuts as opposed to cleaver chops was noted on the hind limb. However, there were indications that the cleaver was also used to disarticulate the femur and tibia. Prep marks were used to locate and expose the ball-and-socket joint between the femur and pelvis. Unlike the practice seen on the Romano-British sites, the medieval assemblages showed evidence that a blade was used to disarticulate and debone this region. Two experimental techniques were carried out to study the pattern of disarticulation at the femoral-tibial joint, replicating two distinct procedures that were noted on the archaeological bone (Fig. 10.7). The first involved the use of blade (point insertion), penetrating the joint at an angle, in order to detach the medial and intermediate patella ligaments. The resulting mark was singular and unexpectedly showed a strong correlation with at least one archaeological specimen, also from a juvenile animal. This cut resulted in the tip inadvertently slipping under the epiphysis, leaving a mark on the metaphyseal surface. This mark would have been inadvertent, as there would be no way to locate the metaphysis with flesh and tendon on the joint. Although this was not replicated precisely, the similarities in terms of cut placement indicate the same underlying function. Overall, these techniques show a greater degree of variability within the butchery process, and perhaps a wider range of tools and flexibility in tool use, for this period.

Axial Splitting

As only one carcass was available, both the Roman and medieval assemblages are summarised here. Replicating the sequence from the Roman materials provided convincing evidence that the carcass had been disarticulated gradually, with the limbs removed prior to splitting of the vertebral column. This conclusion is reached as follows: to create two roughly equal halves,

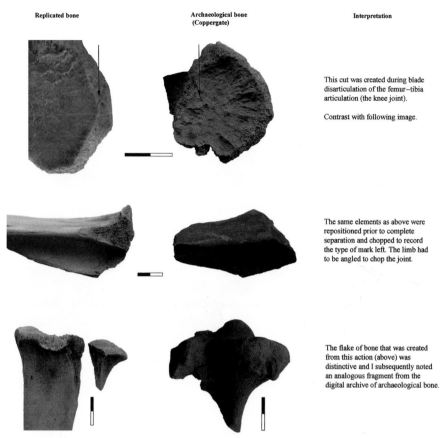

Replicated bone   Archaeological bone (Coppergate)   Interpretation

This cut was created during blade disarticulation of the femur–tibia articulation (the knee joint).

Contrast with following image.

The same elements as above were repositioned prior to complete separation and chopped to record the type of mark left. The limb had to be angled to chop the joint.

The flake of bone that was created from this action (above) was distinctive and I subsequently noted an analogous fragment from the digital archive of archaeological bone.

FIG. 10.7 Hind limb elements.

I split the carcass axially on either side of the spine. This was the most effective means of splitting with a cleaver, and this scenario is corroborated by archaeological examples (images 1 and 2 in Fig. 10.8). This technique results in the pelvis being split at the neck of the ilium. The resulting joint, effectively the hind limb, contained a portion of the pelvis bone, which was subsequently deboned. In this case, the two techniques, axial splitting versus chopping at the femoral head, are mutually exclusive: splitting the carcass precludes the need for further disarticulation of the femur. In contrast, for the medieval period, a wider gamut of techniques was utilised to split the carcass, particularly from Coppergate. However, the underlying driver appears to have been a desire to cleave the carcass into halves and quarters.

CONCLUSIONS

Cirencester provided a rich body of evidence to which I could apply and test the boundaries of the butchery recording protocol outlined in Chapter 9. The fact that it was possible to reveal and distinguish a pattern of butchery on a per

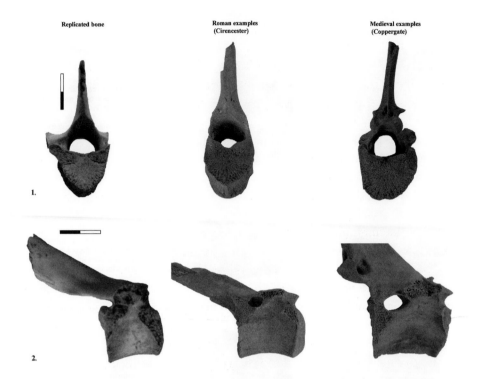

Replicated bone

Roman examples
(Cirencester)

Medieval examples
(Coppergate)

1.

2.

Image 1 (caudal profile) illustrates the outcome of chopping to either side of the spine. Image 2 shows this activity from a side profile. Although examples of this type of splitting were evident from Coppergate (as illustrated here) there are also indications that the spine was divided through the middle of the vertebrae.

3.

Image 3, with corresponding archaeological examples, shows the characteristic marks left on the rib head following splitting to either side of the spine..

FIG. 10.8 Details of mid-line splitting.

element basis from this site was particularly important for the implementation of the methodological approach. This permitted a view of the butchery sequence directly from the material, a much more effective position from which to deduce the *chaîne opératoire*. Analysis of material from Cirencester led to inference that revealed considerable depth, which in turn supports the application of the method and its inherent flexibility.

The case study also emphasised the importance of developing ways to constrain the large body of data that is invariably produced when studying cut marks and distil from this the main features that would be useful for zooarchaeological purposes. This presents a challenge to improving cut mark assessment. While the bone from Cirencester represents a sizable cohort, many assemblages are considerably larger. These points signal the potential benefits for clarifying the economic, social, and cuisine drivers, should zooarchaeologists adopt the method presented here more widely.

The butchery units, in turn, represent a point of resonance with current research. The units have been achieved through robust methods, corroborated through three datasets and validated through experimentation; they could serve as a point of reference for other zooarchaeologists working at similar sites. However, they are not applicable to all situations. Their development is based on the operational sequence interpreted from the cut marks; they are not based on assumptions relating to resource acquisition or drivers such as transport. The crucial component has been to identify the intricately connected association between implements and the cut marks. By combining the interpretation of these two data strands it has been possible to gain a clearer account of both the technical aspects of the butchery processes and the underlying principles that have governed the activity observed from the archaeology.

Differences between the periods were evident. For example, there was an emphasis on cleaver butchery from the Romano-British assemblages, whereas the trend from the medieval sites favours use of various blades for jointing and disarticulation; this distinction was particularly clear when comparing the fore- and hind limbs. The results from the Romano-British assemblages suggest a high occurrence of meat removal. In contrast, the same carcass portions from the medieval sites point to an equally strong bias toward gross disarticulation. The possible reasons for this variability are discussed in the next chapter. Chapter 11 also offers explanations for some of the idiosyncrasies that have been noted in this chapter. For example, there are clear indications that the medieval cleaver was a sharp and versatile tool. It is therefore surprising that it was not used more extensively for disarticulation, particularly when contrasted with the way the Romano-British cleaver was utilized. On medieval sites, the cleaver was apparently reserved for bone breaking and pot-sizing activity. This may have related to aesthetic considerations or the demand for meat without bone splinters, such as roasts.

CHAPTER ELEVEN

CONNECTING COMPLEX BUTCHERING TO COMPLEXITY IN SOCIETY

BUTCHERY AS A LENS ON ROMANO-BRITISH AND MEDIEVAL WORLDS: FROM THE 'GROUND UP'

From an economic standpoint, Roman Britain has been compared to other regions of the North-West Province, with suggestions that hides and leathers were an important export (Fulford 1991: 35–47). More specific commercial implications have also been drawn – the faunal remains from Roman Lincoln, for example, provide convincing evidence for a 'trade in grease' (Dobney 2001). The contribution of Roman butchery to the economy has been based on evidence that points to systematic and large-scale activity, thus a defined meat trade (Maltby 1979, 1989, 1994, 2017; Grant 1989; Dobney 2001). The archaeology indicates that military personnel were instrumental in disseminating new butchery techniques (Grant 1989).

Approached from a sociocultural perspective, the period is also characterised by transitions in religion and demographic flux (Salway 2015: 15–41). Population dynamics might be more appropriately included under economic rather than sociocultural trends. However, for the purposes of this book, more than the number of individuals, it is the diversity of ethnicities, and the ramifications that this may have for understanding identity, which are of principal concern. Given strong connections between identity and cuisine, for example, it is easy to appreciate that regional ethnicity and characteristics must have had an influence on butchery and meat consumption (Hurst, 2005: 303). In short, whether considered from an economic or social viewpoint, the role of butchery as a means of assessing changing attitudes of animals and their bodies may well enable new under-standings of Roman Britain.

A brief review of the medieval period similarly points to a complex, and potentially insightful, human–animal context. Research suggests that changes in husbandry, increased social complexity, and religious influence (Billington 1990; Albarella 1997) had implications for the medieval meat trade and, as a consequence, butchery practice. As commodity and symbol, meat had an important role to play within medieval society (Woolgar 1999: 111). As profession and professionals, butchery and butchers were central to the broader commercial development of later medieval industry and were integral to the day-to-day and festive occurrences of medieval life (Billington 1990). The craft provided a means by which social divisions could be expressed, e.g., through the enactment of representing meat in specific ways during feasts. Anecdotally, we see evidence for the commercial and, by analogy social, importance of meat provisioning in the fact that London actually contained proportionally more meat and fish sellers by the late medieval period than it did in recent history (Billington 1990).

If a gap exists in our research, it is in the fact that we have not been able to tease out adequate detail and nuance in our existing examinations of the archaeological record. Establishing the meaning inherent in butchery, as an enacted craft and embodied activity, would allow archaeologists to better describe past activity. What were the implications of different attitudes to meat consumption? How did identity influence the techniques used to procure and process meat? Thus, in reconsidering butchery 'from the ground up', a number of principles are described below that form focal points of butchery. In turn, these provide insight into the drivers that governed transitions in techniques, tools used to butcher, market systems, and networks of dispersal. How we integrate across these lines of evidence will be critical to our ability to achieve a deeper, richer, and more representative evaluation of past human–animal interactions.

Following a brief summary of evidence drawn mainly from Cirencester and Coppergate, the results are then used to generalise, implicitly relying on cues discussed in Part I. As only cattle are discussed, some caution is needed; other species would be subject to different norms. Pork, in particular, regularly noted as being cured, would have been treated differently. The aim, by focusing on quotidian practice and relatively minor components of the butchery process, is to emphasise the significance of slight variations in technique and technology. Through these technical details, I aim to illustrate the descriptive power of different components of the butchery process, and how these contribute to broader inference.

INFERRING BEHAVIOUR FROM PATTERNS
AND OPERATIONAL SEQUENCES

Three parameters emerged as being of particular significance for interpreting the archaeological context:

1. The underlying principles of butchery
2. The techniques used and whether these could be considered 'period specific'
3. The qualities required and incorporated into the implements and the implications this had for the manner in which they were then employed for carcass processing.

In short, these same points refer to the craft of butchery, underscoring why 'practice' is of seminal importance for understanding the wider subject matter. The following situates the results as they relate to butchers and butchery from the Romano-British and medieval periods and discusses some of the wider ramifications these results may have.

A distinct operational sequence was proposed for each period, represented schematically in Figs. 11.1 and 11.2. These depictions are effectively a summary of the cut marks that indicated 'gross disarticulation' and illustrate the main techniques of disarticulation and the subsequent carcass portions.

The challenging aspect in establishing these operational stages is deciding which marks should be discarded and which retained. How was this done? In this instance I have relied heavily on past experience. In addition, I have engaged with modern butchery manuals that illustrate standard practice, posing the questions: how do contemporary techniques compare with the archaeological evidence? Could these provide clues as to how the ancient practitioners divided the carcasses?

The most useful means of filtering was the dataset itself: which marks allow the analyst to record the 'trinity' of butchery (location and type, the tool and its qualities, and the activity)? The qualitative records, noted on templates and photographed, helped to identify clustering, highlighting areas of the carcass to pay special attention to. Was the ancient butcher driven to process a portion in a specific way? Finally, the analyst can focus on those elements that precedent – from other studies or published materials – has shown to be of particular value for understanding process; e.g., phalanges and metapodials often show evidence of skinning, but ribs are invariably chopped in a routine manner with little idiosyncrasy (Crabtree & Campana 2008: 324).

One final point to note with regard to the schematic representations above: they are a stylised visual depiction of a sequence to illustrate general trends and are themselves based on conflated datasets, which brings a number of limitations (see Chapter 8). Such models may serve to encourage greater accuracy and precision in how we construct our comparative

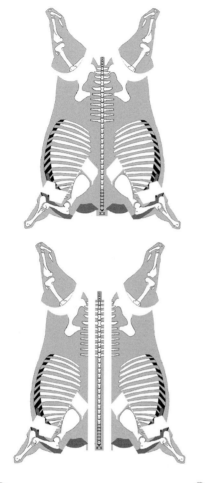

The fore and hind limbs were the first parts to be detached from the body of the animal, leaving the rib cage, vertebrae, and pelvis as a unit. It is possible that the head would have been removed before this juncture, as logistically this would have facilitated subsequent processing. However, there are no archaeological correlates as to the precise sequence for head removal.

After removing the limbs, the carcass was split. This was achieved by one of two ways, either by sectioning the spine transversely first, or with the spine split axially. In either case, sectioning occurred before 'jointing'. Both approaches are observed in the archaeological record.

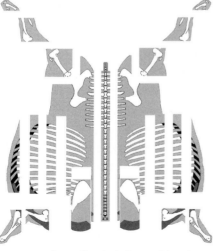

Final processing would have taken place to reduce the carcass into portions suitable for cuisine. This phase of activity incorporated all Stage 3 techniques, such as meat removal, bone breaking, and pot sizing. The cleaver was used extensively throughout this final phase. It would appear that aesthetic considerations were of little importance, as splintering would have occurred with cleaver use. Furthermore, it is possible that all these phases were carried out within a relatively short space of time, perhaps even as one (protracted?) event.

FIG. 11.1 Gross disarticulation and butchery sequence for the Romano-British assemblages, suggesting a systematic approach to processing.

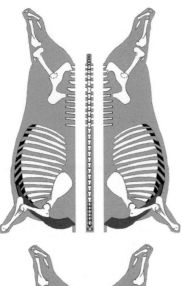

Although greater variability was noted on the medieval sites, it appears that splitting the carcass along the spine was the first major carcass division. This would have resulted in two distinct halves, either with one side retaining most of the vertebral column or with the vertebrae left more or less as a unit as illustrated in this image. The more likely scenario is that the vertebrae were left on one side of the carcass.

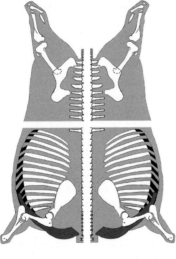

The next major division, transverse sectioning, rendered four quarters. It is worth mentioning that there are indications from Coppergate that the order outlined here, with the first and second diagrams, may have been reversed. This is discussed in greater detail in the main text. It is important to remember that regardless of the sequence, the butchery was performed to produce quarters of beef.

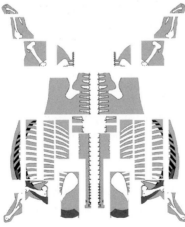

As with the Romano-British sites, final processing would have taken place to reduce the carcass into portions suitable for cuisine. The main difference lies in the fact that mid-to long-term storage was apparently the aim of this mode of butchery and once the carcass was quartered it was likely stored for a period. Aesthetics may have been more significant as blade-use dominates much of the 'subsequent butchery'. The cleaver was reserved for bone breaking, apparently after the meat was removed.

FIG. 11.2 Gross disarticulation and butchery sequence for the medieval assemblages, which points to more standardised practices.

frameworks. Ultimately, such comparisons will become more robust in their ability to inform us about the complex societies we study. However, these need to be tested and enhanced. At present, the type of interpretation they provide has limited utility if our aim is to capture *nuance*. These representations would become more useful, and usable, if we recorded cut marks to a more standard format. Only then can these models contribute to the regional modelling and synthesis that is of utility to zooarchaeology (King 1999; MacKinnon 2014).

In practical terms, the distinction between 'systematic' and 'standardised' (Figs. 11.1 and 11.2) rests on the evidence that suggests a piecemeal approach, such as chopping the femoral head for limb removal (Roman Cirencester, for example), versus the quartering style of butchery from Coppergate (medieval).

In this instance, 'systematic butchery' implies that the same techniques are being used, but the outcomes are less specific; i.e., the result is not a specific 'cut of beef' as in the modern context, it is a 'piece of meat'. This would appear contradictory in that if a butcher is using the same methods to cut, surely he or she will produce the same portion of meat? The point that needs clarification is that the butchery is systematic with regard to the type of actions and implements involved (predominantly cleaver chopping), but the outcome has less to do with the aesthetics of the meat.

For the medieval period, 'standardised' might again seem contradictory, considering the level of variability observed on material from Coppergate (O'Connor 1989b) and the dissimilarity in butchery activity and implement use noted from the medieval sites.[1] Here 'standardised' implies that a number of different cutting techniques are used, along with a variety of tools, but the outcome is the same, resulting in a carcass that is quartered, for example. How this is achieved is less important than the outcome. The main concern is the production of the side, quarter, or joint of beef. In this scenario, aesthetics and the joint itself would appear to be of greater importance.[2]

Insights on Storage

Delving more deeply into the implications of different strategies of butchery, during the Romano-British period we see evidence for an approach commensurate more with immediate consumption and/or further processing and less with a view to longer-term storage. As discussed in Chapter 5, from a commercial perspective fresh meat that is to be stored is generally left in as large a portion as possible, regardless of whether the portion will be refrigerated.[3] Thus, longer-term storage of more than a few days necessitates that the carcass be left in a substantially large portion, such as a quarter (Romans et al. 1994: 546). This also serves as an important division of the carcass that guides subsequent meat cutting (Gerrard 1964: 265–77).

By removing the limbs, a greater area of cut flesh is exposed, leaving the meat prone to spoiling. If cut into smaller units, without refrigeration, the flesh has to be salted or smoked. In both scenarios, the butchery process remains the same; the distinction lies in whether the carcass is stored as opposed to preserving the meat.

Indications that the limbs were removed individually would corroborate the hypothesis that the Romano-British butchers were processing the whole carcass in a single event. Once the limbs were disarticulated, the remaining axial components could then be portioned into smaller sections. In this scenario, it is unlikely that the carcass was stored. Quartering the carcass, as suggested from the medieval assemblages, indicates the opposite requirement: the carcass may have initially been retained in large units, and storage was an important stage within the butchery process.

What Are the Implications of More Immediate Use versus Longer-Term Storage?

What might be some of the expectations of storage? If meat was smoked or salted, when would this occur? Tackling these questions should not distract from the fact that we are observing different traditions of practice from the two periods. Fundamentally different ideals and principles guided the butchery. By cutting the carcass into sides of beef, as noted by O'Connor (1989b: 154) from Coppergate, or quarters as indicated from this study, the evidence would suggest that the medieval butchers were stowing large portions of the carcass. On the one hand, this has commercial implications, by providing better management of supply. With several carcasses hung in a relatively cool setting, the butcher can better modulate supply and anticipate demand.

There are also important implications for cuisine: could this activity relate to meat quality, in this case, the desire to 'mature' the beef? Maturing beef by suspending the carcass once it has been skinned and eviscerated is carried out to improve taste, texture, and flavour (Gerrard 1964: 237–40). Although subjective, the perceived improvements in the quality of the flesh have meant that maturing to some degree has long since been a part of the meat industry (Romans et al. 1994: 548; Sitz et al. 2006). A desire to mature meat offers another reason, which complements the suggestion above of anticipating demand, for why the carcasses were being quartered. Could this speculated desire to mature the flesh also provide deeper insight into the perception of meat for those in society who were able to express specific tastes? Referring back to the Romano-British period, perhaps we are observing evidence for a less developed 'meat palate'. This may be as a consequence of meat itself only being an occasional food for the majority of inhabitants, alongside the influence of Roman ideology toward the diet: meat as a constituent, rather than the main constituent.

By the medieval period, we have good evidence for the fact that meat was eaten in relatively larger quantities (Woolgar 2006: 90), which may have galvanised diversification of meat qualities, increasing appreciation for different textures and flavours. This introduces numerous dimensions connecting the techniques of butchery to society at large. It is this type of nuanced interpretation that helps to tackle some of the wider questions: given similar population sizes and institutions (comparing the mid- to late Roman period with the late Middle Ages) what are the reasons for differences in provisioning of meat? How does this connect to other foodstuffs? Clearly, there are the different ethnic groups and ideologies involved – underpinning social variability – but also the military influence associated with Roman occupation did not exist during the medieval period. How do these larger institutions come to influence day-to-day life?

Materiality Reveals Practice

Implements are a useful archaeological indicator of material objects that are specific to a given craft. Distinctive knives, including cleavers, have been recovered with regularity from Romano-British sites (Manning 1966, 1976, 1985). Trade equipment also lends itself to helping decipher how the craft progressed. For example, meat hooks have been recovered from London, Silchester, and Verulamium (Mac Mahon 2005: 63). The three-legged chopping block, typical of the Roman period, was still in use at least up to 1600–1700 (Rixson 2000: 72), and a larger table-sized version continues to feature in the modern butcher's shop. A utilitarian item such as the three-legged block may have been popular simply because of its functionality. However, this does not preclude such items from becoming associated with certain professions. The steelyard scale, another Roman introduction to Britain, has been excavated throughout the Roman Empire (see bronze examples mentioned in Thomas 1963, although these relate to medical instrumentation). Taken together, these three objects, hooks, block, and scales, alongside the range of knives, illustrate craft specialisation. Given the fact that these particular items apparently endured and were used well after the Roman period, could they have been a feature of the developing professionalisation of butchering? Furthermore, this type of standardisation of craft objects would probably only have occurred as a consequence of the craft itself becoming a trade in a stricter sense, encouraging the routine production of these items.

Beyond serving as symbolic and material markers of specialisation, these objects have much more nuanced meaning. The butcher's block is far more than a convenient lump of wood on three legs. The block elevates the carcass, metaphorically and literally, as well as the activity of butchery. What would previously have taken place on the ground (see Fig. 6.5) is altered in ways that fundamentally change the way the craft is performed. The techniques transition from those designed to facilitate butchery whilst the butcher is kneeling, squatting, or leaning over to ones designed to take advantage of the upright stance that is more natural to human anatomy. What are the implications of this relatively small development on the tools and techniques to process meat, and the physicality and performance of the butcher? For one, the cleaver is a type of blade best suited for use against a block. The block provides a solid platform to maximise efficiency while chopping, one of the functions for which the cleaver is designed. Thus, the cleaver and block are complementary.

Furthermore, the cleaver can be used from a higher vantage point and engages the levers of the wrist, elbow, and shoulder. The full momentum of the body, whether by slight flexion of the leg muscles or more engaged involvement of the upper-body muscles and flexion of the back, can be brought to bear. In effect, the cleaver can be used more effectively, and the technique of chopping can become a more prominent feature of butchery, as can the tool itself. Viewed from this perspective, one hypothesis for why butchery in the Roman period is fundamentally different from that in the Iron Age is that not only are new tools used, but the essential *locus* also changes, elevated from ground level to the butcher's block. Therefore, being able to decipher whether or not a block was used has many important ramifications.

What do the knives, and their mode of use, reveal about butchery during these two historic periods? An intriguing aspect of urban Romano-British butchery practice was the use of tools and techniques that sped up carcass processing, compared with earlier Iron Age traditions. Chopped femoral heads, scoop marks, and the particular techniques used to process the scapula all point to rapid methods of portioning. These techniques were used in favour of those that would have required less physical effort, but longer cutting times.

To illustrate this point, the butchery sequence for dismembering the femur, using the Romano-British cleaver, requires no more than one or two slicing cuts into the muscle of the thigh followed by a single chopping cut to release the femur. At most, two further slicing cuts would be needed to completely separate the entire limb from the carcass. In contrast, using a knife would require several slicing cuts to deal with the tough ligaments and tendons surrounding the hip joint, and then to release the leg. In terms of time allocated, the latter technique would require at least three times longer to execute. The preference of the cleaver over the knife also supports the conclusion that urban and military Romano-British butchers were trying to reduce the amount of time spent processing *each* carcass. However, swift processing is not necessarily indicative of immediate consumption. The beef may have been brined, salted, or smoked, as suggested from Lincoln (Dobney et al. 1996: 26) and recently from sites in southern England, for example

(Maltby 2017: 203). As noted above, preservation versus storage of meat drive butchery in different ways.

For the Romano-British period, chopping, 'prep-chop', and 'slice-chop' marks all depend on a curved cleaver. The Roman designed cleaver is versatile and used for chopping, slicing, and perhaps meat-paring activity. Could this tool represent the point at which, at least for Britain, specialist butchery tools are first created? The implication here is that these tools are not knives used by a range of craftspeople for different tasks but retained and used only for butchery. Certainly, by the medieval period, good quality specialist implements from cleavers to a variety of smaller blades had emerged and were being used with consistency by professional (urban?) butchers.

Dealing with Demand

In a commercial context, storage and speed of processing revolve around the ability to suspend the carcass. In a modern setting where the carcass is hung, split longitudinally down the midline, and then quartered, the whole process of dismemberment is facilitated by specialised industrial technology. However, without the advantages of automated lifting mechanisms, the Romano-British and medieval butchers would have had to rely on simple pulley systems to hoist the carcass (Fig. 11.3) or carry out the dismemberment process while the carcass was lying on the floor.

Hanging a carcass necessitates vertical space but saves a considerable amount of ground space. However, hoisting and suspending a large animal requires a number of additional considerations. It is possible that specific buildings were created, or parts of premises converted, with tiered platforms to accommodate the needs of a butcher processing a large carcass (Gerrard 1964: 97–8). Evisceration and skinning are both facilitated if the carcass can be suspended (Fig. 11.3), as gravity assists with removal of the internal organs, which can be deposited into a waiting container (FAO 1991). If the animal is to be eviscerated at ground level, the butcher must remove the entrails and internal organs manually.

In effect, once the carcass is suspended, a single person can undertake skinning and dismemberment, as opposed to between two and four individuals that would be required to process a large animal while on the ground. Texts describing home butchery suggest that at least one and preferably two to three assistants are needed to help when butchering a cow (Mettler 1987: 9). The additional aides help to turn the carcass while a third person flays the skin. As well as facilitating evisceration and skinning, suspending a carcass also leaves the butcher more options for storage and when to culminate the butchery process. If an animal was slaughtered, gutted, and skinned on the ground, it would need to be dismembered at the same time. Leaving a carcass lying either

FIG. 11.3 Suspending apparatus as used in a small commercial enterprise, Kimana slab, Kenya.

on the side or back would cause blood to accumulate in the dependent parts, rendering the meat unfit for consumption. Furthermore, this would leave the carcass prone to infiltration by bacteria (FAO 1991) and increase the likelihood of insect or rodent infestation. By suspending the carcass, the butcher can leave the carcass for days or even weeks before further processing; indeed, this is essential if maturing meat. Thus, in addition to manpower, time becomes a critical variable if the carcass is not or cannot be hung.

These practical difficulties would have been pertinent during the Romano-British and medieval periods. Archaeological evidence points to hanging as part of the initial butchery sequence, for skinning and evisceration; this is reinforced by literature describing non-mechanised butchery (Mettler 1987: 9; Maltby 1989: 12–14; Hambleton & Maltby 2009). However, such a scenario is possible only in specific circumstances, for example in purpose-built slaughterhouses or where space permits hoisting equipment to be erected.

Drawing some of these strands of evidence together, the findings from sites such as Coppergate may also have commercial implications. Borrowing from the modern context, the method of processing into quarters is similar to contemporary systems where beef carcasses are sold in wholesale markets as halves or quarters and subsequently processed into smaller portions, 'meat cuts' in the modern sense, on the butcher's premises (Ashbrook 1955: 11; FAO 1991; Swatland 2000). This reduces the likelihood of spoilage, as a quarter-carcass is suitably intact to prevent premature deterioration.

There are some practical considerations to take into account. Dispersal of meat from markets and wholesalers to a butcher's premises, as opposed to large retail outlets where mechanical aids are easier to use, depends on the butcher's physical ability to transport the carcass. Virtually all domesticates are purchased as entire (or half) carcasses, which have been dressed. Further processing takes place at the butcher's shop. Cattle are an exception. A quarter-carcass is the largest portion size that can be comfortably transported by one person, for example from a loading bay to a van, for transit to the butcher's premises. Today, if a joint of beef is processed prior to arriving at the butcher's shop it is invariably vacuum packed to prevent spoilage, or 'wet-matured' (Sitz et al. 2006).[4] Furthermore, it is worth noting that alongside the archaeological evidence, historical sources from London point to the practice of quartering. In 1556, the Wardens of the Butchers' Guild promised the alderman that when a customer purchased 'a quarter of beef or a whole carcass of mutton it would be delivered by one of the butcher's own servants, and not one of the idle or poor people standing around the shambles' (Jones 1976: 139). This was implemented to reduce fraud and deceit for those customers making a substantial purchase of meat. Indirectly, it suggests that a quarter-carcass was the largest unit in which beef was stored on the butcher's premises. This idea is bolstered by an account from Goodrich Castle, which highlights that the underlying butchery techniques themselves for processing beef were directed toward preservation and conservation of meat for mid- to long-term provisioning (Woolgar 1999: 116).

These points would support the notion that medieval butchers were procuring meat as quarters. Could the material from Coppergate thus provide evidence pointing to the residue of butchers' waste in the stricter sense, butchers functioning in a semi-professionalised state serving as intermediaries between farmers or wholesalers and the public? It should be made clear that the materials from Coppergate (eleventh to twelfth century) derive from a period before the York Butchers' Guild is established, which occurs in 1272, gaining momentum in 1299, after which there is an almost annual entry of butchers into the Freeman's Roll (Corsair & Fitzell 1975: 1). Thus, the implication is not 'guild-type' professionalisation, but perhaps the precursor to more standardised practices that endured until the development of the abattoir system.

In summary, a critical point signalled in Chapters 5 and 6 is that relatively minor technical aspects of processing can have far-reaching implications. Not only do we observe subtle changes in technique, which modulates the overall sequence of butchery dramatically, but we can also appreciate the influences on building design and organization of the taskscape. These features serve as indirect evidence for levels of demand required to accommodate the intensification in trade. Comparing the two traditions of butchery, the piecemeal approach from the Romano-British period involves less complexity than that observed in the medieval period. For the latter period, the research suggests longer-term storage of fresh meat. However, this distinction should be considered within the context of a more diverse and complicated butchery system. As with any setting, cuisine and a range of socioeconomic factors must have influenced medieval butchery. For the larger urban enclaves, at least, a degree of standardisation may have played an equally important role. Needless to say, many approaches to butchery would have been in place, particularly when taken within the context of 'medieval Britain'. To illustrate this point, historical records show clear evidence for both fresh and cured meats. In 1405, Sir Hugh Luttrell brought animals from Wales to his house in Dunster, some of which were shipped via the Bristol Channel and salted down (Woolgar 1999: 113); the Duke of Buckingham, in 1501, hired a slaughterhouse and butcher from Southwark to supply his house on Queenhithe with fresh beef and mutton (Woolgar 1999: 114). Though in distinctly different contexts, and nearly a century apart, the above examples serve to illustrate the obvious complexity involved not only in the processing of meat but in its procurement and transportation. In turn, this evident complexity serves to illustrate the work ahead of zooarchaeologists if we are to capture, assess, and integrate nuance into our study of butchering.

Scales of Activity

At this stage, I would like to step away from the archaeo-historic context and once again turn to ethnographic research as a way to add nuance. Rather than drawing inference from the archaeology, as it may relate to the development of the Roman meat trade or professionalisation of butchery practice during the medieval period, I will instead use a modern example that may help to situate the archaeological case. Building on the vignettes presented in Chapters 3, 5, and 6, here I expand the Kenyan example to illustrate one route through which changes in practice may be initiated and institutionalised.

The modern meat trade in Kenya is in part a consequence of British imperialism, which established the East Africa Protectorate in 1895, subsequently known as Kenya Colony from 1920. An independent Republic of Kenya emerged in 1964. As a consequence of European imperial rule, new systems of land management and agro-pastoralism were instituted (Morgan 1963). However, perhaps the most far-reaching impacts were as a consequence of radically different ideologies dealing with currency, 'ownership' of land, and social stratification, both during and after the period of colonialism (Maxon 1992a: 249–58; 1992b: 273–96; Ocheing 1992: 259–72; Fiona & Mackenzie 2000). This had the effect of disrupting long-held systems of mobility across the landscape, many of which involved the movement of animals. With the British, new systems of governance and legislations were also instituted. This included the formation of the Kenya Meat Commission, established in 1950 by an act of parliament. The act was principally concerned with providing a market for farmers and quality meat for consumers. The Meat Control Act was set out shortly after independence, in 1972, primarily as a mechanism for quality control and to institute best practice in slaughterhouses. In short, with colonisation new ideologies surrounding meat provisioning were rapidly institutionalised and continued to be the norm after independence, despite the relatively short colonial period.

Although ostensibly the historic references to the Kenya Meat Commission and the Meat Control Act provide evidence of European ideals of how meat should be produced and sold, the lived experience is understandably more complex and revealing. These legal measures set out to promote and monitor growth at a time soon after new land borders had been established or were still in flux (Ogonda & Ochieng 1992: 1). In some ways, they responded to modernity and sought to promote national and international export. However, for many parts of Kenya the realities of the movement, marketing, and slaughter of animals have changed little. Alongside this stasis, new systems have emerged that now function in tandem with traditional practices, themselves meeting different and new forms of demand.

For the region in which I worked, Kimana and the surrounding area (Fig. 3.1a, Chapter 3), cattle, sheep, and goats are still moved around the landscape in a way that mimics ancient practices, if not along the same ancient routes. This region is predominantly constituted of land owned by the Maasai, a group who adhere strongly to the ideals of pastoralism and vehemently

FIG. 11.4 Domestic animals being moved from one market to the next along existing roads; stopping off points allow animals to rest. For the Kimana region, livestock are moved some 90 kilometres from Kimana to Emali, taking three to four days to complete.

eschew agriculture (Århem 1987: 3). With the increase in infrastructure, in particular, roads, new towns have sprung up along these highways. Alongside this incipient urbanisation process, the roads are also now used as major thoroughfares for moving cattle, which one can observe almost as soon as one leaves the capital, Nairobi (Fig. 11.4).

For the region surrounding Kimana, domestic animals are moved along a route of some 90 kilometres, from one local market to the next (Fig. 11.5). Following market day, Tuesday in this case, animals that have been purchased are slaughtered on Wednesday in the commercial context, or, if purchased for later consumption, the animals are then transported to the homestead on the hoof.

Recapping from Chapter 5, slaughter takes place in four distinct settings in Kenya: around the homestead or in one of three commercial establishments, a 'slab', slaughterhouse, or abattoir. In the Maasai homestead setting, for example (see Chapter 6), animals are butchered to defined standards governed by social norms that centre on who receives which part of the carcass (Seetah in prep b). Cattle are only butchered for feasting occasions, not for 'food' (Århem 1987: 9). Maasai learn to butcher in childhood. Slaughter and butchery is defined along age and gender lines, but in general the tasks fall into the following order during a feast: men kill, butcher, and prepare their meat;

FIG. 11.5 Market day in Kimana. This falls on a Tuesday, with slaughter taking place on Wednesday at the local slab.

women will clean entrails and prepare their own meat; and young boys, for example, will crack phalanges, as the distal limbs are assigned to this group. As the example describing Daniel Mabuvve's butchery indicate, choice of cutting tools is not particularly important (Fig. 6.5, Chapter 6).

In contrast, butchery at the slab and slaughterhouse are dictated by commercial practice. Both employ similar techniques but are differentiated based on scale of activity and size of establishment, the slaughterhouse being the larger enterprise. Using the Kimana slab as a case in point (see Chapters 3 and 5), slaughter and butchery takes place on Wednesday, following market day. If a large number of animals are to be butchered, one session of butchery takes place during the early to mid-morning, and another in late evening. The system is organised with areas to corral animals to be slaughtered, dedicated waste management systems, and mechanical pulleys to assist with butchery (which may be mechanised in the slaughterhouse). Specialist tools are used, and taskscapes are well defined, with distinct divisions of labour. In the three slabs I visited, all tasks are performed by men. Cattle carcasses were halved and then quartered to facilitate transport to retail establishments. Modes of transporting meat ranged from small trucks to specially marked delivery trailers to motorbikes (Fig. 11.6). In addition to waste control, quality control was also a key feature, with meticulous veterinary inspection taking place at intervals throughout the day and night. Following butchery, retail takes place from butchers' shops, which are small and numerous; in the growing urban areas, I estimated at least four butchers per square kilometre.

FIG. 11.6 Small-scale transporting systems for dispersal of meat to local butchers' shops.

The example from Kimana illustrates a system where many animals are butchered in a short space of time – a single day – based on commercial drivers. This mode contrasts with homestead butchery, where one animal is butchered for immediate consumption, usually within a ritualistic context, and subject to strong cultural norms. In the homestead case, butchery is piecemeal; the order of removal of individual parts does take into account the animal's morphology, but generally, detachment is driven by the need to provide specific portions that are then shared with members of the community to a defined schema. The skills and the techniques of butchery are formulated around this arrangement.

In the small-sized commercial context of the slab, one owner employs many individuals who work intensely one day (or, if demand requires, two days) per week, but who also have other jobs. Part-time specialists undertake the butchering and ancillary tasks, such as removing the final slivers of flesh from the skins or washing entrails. The butchery is regularised, heavily governed by economic dimensions, and follows a pattern that feeds into systems of transport and retail, not social ideology. The skills are learnt on the job and, while consistent, do not encompass the scope of divisions into 'cuts' as observed in Western meat retail. This is because no such desire for these types of cuts exist, nor does the animals' morphology lend itself to differentiated musculature. I observed little evidence of cattle being used for traction; most were clearly raised exclusively for consumption. Further, the animals are well maintained and much discussion takes place during market day as to what constitutes a 'good' specimen, especially for cattle. However, the animals travel long distances across the landscape between markets and are often mature. Thus, in this region at least, meat is meat, with little preference expressed for different parts of the animal in everyday settings. Contemporary day-to-day cuisine, even for the Maasai, who place great store on meat consumption, integrates meat as part of the meal; only during ritual feasting is meat the main component.

The approach used in the slab modulates many factors. For example, it provides time for waste to dry, ultimately causing fewer nuisances within the local environment. The scales of processing have changed, from homestead to commercial, as a consequence of changing availability and attitudes to meat, but also because more fundamental transitions have taken place that influence ideology, governance, land organisation, and human–human/human–animal interactions.

This vignette is not presented to imply direct correlation between the archaeological and modern, but rather to illustrate a number of features that are not clearly visible from the archaeological context. The patterns of processing observed in different settings help explain some of the distinctions observed in the Roman and medieval periods. Even though there is strong evidence to suggest commercial practices were the norm in the Roman period, the piecemeal approach to butchery may reflect cultural ideals including cuisine, when meat is simply meat. The more standardised practices, halving and quartering, as observed in the small-scale commercial context of the slab, may similarly offer a lens onto the ways in which increasing demand mandates the development of specific techniques to meet intensified consumption. The enacted tasks, grouped and defined into taskscapes in the slab, offer a view of the variability that was likely evident in the archaeological case. Some individuals work as slaughterers and butchers in the stricter sense, while others occupy ancillary roles. The transition from part-time, arguably semi-skilled, to full-time, skilled labour no doubt forms an important feature of the process toward professionalisation.

Connecting technical features from modern and archaeological settings is useful in describing activity and the tasks of individuals. However, the more important contribution from analogy lies in the ability to situate ideology. At the level of state control, the Meat Control Act is an iteration of one of the measures that fell within the purview of the late medieval guilds. The socio-economic dimensions are wholly different, but the development of greater legislative control is enacted in response to quality control. In the Kenyan case, it is a mid-twentieth-century version of European legislation for managing slaughter that influenced the meat production system.

CONCLUSIONS

In concluding the case study, the results point to diachronic and regional variation, transitions in techniques and modes of processing, as well as wider economic and cultural implications.

Chapter 10 illustrated a number of ways to assess the data from cut marks and distil from this the important features that would be useful for zooarchaeological purposes. Thus, overall, the case study has shown how the methodological approach can capture and constrain the large, complex datasets that comprise the butchery record, and from this produce innovative ways to present the data. The protocol itself can be integrated into existing recording systems and manipulated to fit the remit of other analysts working within different spatial and chronological situations.

The most important aspect of the research is not the recording system but the process of analysis. I paid scant attention to the raw data. Instead, I focused on those details that I deemed most important: the interpretative parameters and outcomes. With the approach presented in this book, the raw data are available, but so too is the opportunity and ability to move rapidly beyond this to the interpretational evidence. The reason for this rests in the fact that these interpretational characteristics form part of the recording process from the outset. There is no ellipsis. Bypassing the raw data effectively illustrates the point of this approach: the analyst need not focus on the minutiae of the cut mark, but can devote their attention to the meaning, and the craft, on which butchery rests.

CHAPTER TWELVE

A CONSTELLATION OF ACTIVITY
EMBEDDED IN SOCIETY

It might be argued that butchering animal carcasses is the single greatest taphonomic (and biostratinomic) factor in the formation of humanly created fossil assemblages.

(Lyman 1994: 294)

INTRODUCTION

Archaeology deals in scales. On the one hand, a narrative about transitions in human society can be inferred from large-scale monuments – for example, by examining physical structures associated with religious practice (Miles 2016: 353). Such architectural features effectively concretise belief in the landscape. On the other, the vast majority of our material evidence actually describes small-scale operations, representing evidence for changes taking place on a day-to-day basis, in households, shops, and between buildings, those spaces where quotidian practice can be enacted. Through varied and heterogeneous processes, these daily activities are then gathered into a specific place, usually a waste dump, which we subsequently study as the archaeological record. Whether grand or small scale, the material evidence often and consistently points to intentionality and the behavioural basis for producing, making, *creating*.

When it comes to butchering, throughout the course of becoming part of the lithosphere, the residual traces of human cognition remain on the recovered materials themselves. What is more, cognition also 'surrounds' cut marks, offering an opportunity to situate signatures of human actions within the wider frame of the practices that led to their creation. It is recognition of this fact – that traces of human cognition remain on the materials themselves – that marks the crucial conceptual change necessary to truly capitalise on the power of butchery. Taphonomy certainly influences the sovereignty of the cut mark, but that should not distract us from our goal of teasing out the behaviour that resulted in the presence of marks.

Current approaches have taken us only so far in our exploitation of the butchery record. The time is ripe to reconsider our expectations not only of cut marks but also of *butchery*. Better methods to assess cut marks, and more broadly to develop a theory of butchery, are key to taking advantage of this rich and illuminating archaeological resource in the future. Butchery, as technology, is an important topic in its own right, one that deserves the kind of focused theorisation and methodological innovation described throughout these chapters.

However, as the arguments of this book make abundantly clear, butchery reaches its full potential as a resource for archaeological analysis only when it is conceptualised as activity enacted and embedded in society. In this guise, it serves as a lens on a range of other subject areas, all of which are connected to the knowledge, skills, and practice of the craft. Butchery is key to understanding the most enduring, complex, and important human–animal interaction: meat consumption. The dynamic relationships we have had with one of our most important foods are highlighted by the archaeological case studies from Chapters 10 and 11; meat had distinct, and different, meanings in each period. Beyond meat and its importance for human social practice across millennia, the study of butchery acts as a revealing window onto activity and craft in the past, the interactions between dead animals and humans, and the relationships between people and 'sentient objects'. In all these cases, butchery is the mechanism through which humans transform animal bodies into vehicles for expressing cumulative economic and social complexity.

Drawing on arguments made in earlier chapters, in this conclusion I briefly explore the potential scope and wider influence of butchery, positioning the ways in which the knowledge and intentionality of butchery may have utility for understanding ideology, people, animals, and objects. I then discuss opportunities for improving our ability to study crafts in the past. Finally, armed with an understanding of the wider influence of butchery alongside recommendations for studying practice in the future, I offer suggestions for how archaeologists might better contribute to current concerns regarding the ecological ramifications of contemporary patterns of meat consumption.

Toward a Global View of Butchery

With the following examples, focused on health, non-food domesticates, the role of gender, and the power of objects to reveal large-scale transitions, I have deliberately chosen topics that are peripheral to butchery per se. However, they illustrate the immensely intricate relationships around the activity itself and the place that butchery occupies in society and, indeed, human psychology. Ethnography and ethnoarchaeology have brought to our attention some of the diversity of roles that butchery has played in human culture. Zooarchaeologists, lithic specialists, and metallurgists have highlighted the economic and technological contexts and, to a lesser extent, sometimes incorporated other cultural factors. This wealth and variety of evidence has much to offer in terms of helping to describe the universality of butchery.

Throughout this book I have emphasised the importance of ideology as a mechanism for driving practice. One such driving force concerns health. As Chapters 4–6 and 11 demonstrate, the relationship between butchery and health has many dimensions. Well-being, broadly defined, played an important role in shaping the wider context of carcass processing. Meat was considered essential for good health, and special licences were granted in Britain to allow butchers to sell meat to convalescents even on days when Christians traditionally abstained (Jones 1976: 124). In contrast, the peril of contracting illness from noxious odours was a fear for the population of London during the late medieval period, at a time when miasma was the dominant theory of disease transmission (Perren 2008: 124). A more overt risk was alleged directly from butchers selling diseased meat during the same period in Britain, which resulted in severe punishment for the offending butchers (Jones 1976: 132).

These historical insights provide a glimpse into a more pressing concern: zoonosis. Raising animals, butchering carcasses, and eating meat exposes us to disease (Horwitz & Smith 2000; Greenlee & Greenlee 2015). One of the more relevant cases is also one of the most recent. Rift Valley fever virus was unknown before its discovery in 1931 in Kenya (Daubney et al. 1931; Arzt et al. 2010; Davis 2010). A major route to exposure derives directly from handling contaminated viscera during slaughter and butchery. In parts of the world where Rift Valley fever virus is endemic, the disease is gendered, due to the different roles that men and women undertake during carcass processing (Muga et al. 2015). Rift Valley fever virus represents one of a host of diseases that are quite literally specific to 'butchers' (Sahay et al. 2010).

The study of the human agents who butcher and how carcass processing impacts those individuals is a vastly underexplored, yet deeply nuanced, topic. Figure 12.1 illustrates features of the butchery ecosystem that deserve closer attention. Evidently, at least two men and two women would appear to have been involved in the butchery of the cow. The objects pictured in this work of art include a man with a large mace, one end of which seems to have a punch-point, probably used for poleaxing; the animal's skull appears to have a corresponding, darker red puncture to the top of the cranium. The woman in the foreground uses a knife to clean internal organs, bent over a three-

FIG. 12.1 David Teniers the Younger, *Butcher Shop*, 1642. (Photograph © Museum of Fine Arts, Boston.)

legged butcher's block. The roles of individuals, men and women, appear clearly defined. Despite its name, 'Butcher Shop', it is not clear whether this immortalised episode of butchery is in fact taking place in an individual's home. Adding layers of meaning through artistic interpretation, the presence of a sieve to the left of the carcass would appear to be a metaphorical reference to *Sapientia* (wisdom), potentially indicating approval of the work of the woman in the foreground (Sutton et al. 1993: 423).

As the Kenyan case studies have outlined, in non-commercial settings the role of women (and indeed children) is much more conspicuous. Centralisation of slaughter and processing had a major impact on those involved in the trade, and how they were viewed by society, as noted from cases in the United Kingdom, the United States, China, and India (Metcalfe 2012: 113; Ahmed 2013; Pacyga 2015: 63; Wang & Pendelbury 2016). When we consider gender and age, there was probably an attendant shift from domestic and feminine to artisanal, public, and masculine, as the trade became increasingly commercial. This is not to imply that women and children did not play an important role; rather, as shown by the Kimana slab example (Chapter 11), commercialisation alters the taskscape in specific and important ways. For archaeology, butchery has tended to be seen as 'specialist', and that is probably the case at least in urban and military settings in the Roman period and commercial contexts from the later medieval period. This does not mean we should lose sight of the varied roles played by different members of a given community.

There can be little doubt that butchery is a topic of central importance to assessing the human–animal relationship in the past. However, the scope of influence extends well beyond the immediate context of slaughtering the live animal and portioning the dead carcass. As discussed in Chapter 5, ideology in the form of new attitudes toward meat consumption led to new techniques to produce specific modern cuts of meat. In turn, this catalysed, and then came to depend on, farmers breeding livestock with differentiated soft tissues. However, other domestic animals were also swept into the ebb and flow of activity around production, consumption, and practice. In much the same way that farmers and hunters bred sheepdogs and hounds, butchers develop dogs to assist their tasks, including for droving and protecting livestock being moved on the hoof. Throughout Europe, a range of breeds have been bred to meet one or both of these tasks: the Rottweiler (Rottweiler Metzgerhund: Rottweiler Butchers' Dog) from Germany (Schänzle 1981), Cane de Brano/Cane Corso, also known as Cane di'Macellaio (Macellaio: butcher, flesher, slaughterer), and the Sicilian Vucciriscu from Italy. Another branch of breed specialisation led to the creation of a variety of bulldog-type dogs throughout Europe for baiting bulls. Butchers in medieval Britain saw baiting as a necessity, prior to slaughter, to make the meat more palatable (MacGregor 2012: 137).

Finally, the spotlight falls on objects. The butchery record provides a lens onto significant socioeconomic changes taking place at critical points in human history. The transition to metal is one such example of a singular, cyclical paradigm shift. Tools that had served their function for some 2.5 million years were suddenly and rapidly (relatively speaking) repackaged into a form that allowed for a wholly different pattern of resource exploitation to take place. This is far from trivial. By connecting the details of increasing implement sophistication with transitions in animal processing, we have an opportunity to investigate changes taking place in society, but viewed from different and new perspectives. Roman iron working was not so much an innovation as a concerted effort at reorganisation. The best techniques of metal working from the Romanised world were marshalled and then disseminated throughout the Empire. This wave of rationalisation influenced the Roman as well as Iron Age groups residing in the periphery of the empire's territory (Tylecote 1992: 62). What was the relationship between production techniques of knives, new types of knives, intensified husbandry, new techniques for butchery, and society?

Once we extricate butchery and butchering – the activity and behaviour – from the products of carcass processing, one of which is cut marks, we reveal an entirely new subject, one which remains largely unexplored as a mechanism for understanding a range of topics – from disease to dog breeds to practice and cognition. The social context is critical. Cultures around the world have articulated social stratification through a range of actions that are centralised around the butchery of carcasses, whether in reference to those who butcher or to the division and distribution of meat. Butchery animates a cascade of influences in human society.

A FUTURE FOR BUTCHERING STUDIES

The above represents a new conceptualisation of butchery, one that views the subject as incorporating a diversity of tasks, with discrete constellations of activity and interrelationships between human and natural agents: butchery perceived as an ecosystem. How do we recover this detail from the archaeology? Answering this question will require more than merely an overhaul of methods; fundamental changes are needed.

Archaeology is ideally positioned to be at the forefront of new insights into this subject. The tools to study butchery from the perspective of technology and agency already reside in the discipline. We can readily acknowledge that butchery provides a useful lens on basic aspects of carcass dismemberment, levels of faunal exploitation, interactions between crafts, and details of organisation in those contexts where a meat trade existed. These features will differ depending on the society in question, and this provides opportunities for comparative studies. However, by utilising the range of new approaches to study 'activity', we could significantly broaden the types of research to which butchery data can be applied. We need an assessment of the linked context between *chaînes* taking place in different spheres of society, incorporating different types of practice and interconnected skills (e.g., the flint knapper as carpenter; see Walls 2016), as well as an assessment of the sociopolitical and economic meaning behind activity. An ecosystem approach also allows for the feedback mechanisms to be included, creating 'integrated *webs* weaving skill, knowledge, dexterity, values, functional needs and goals, attitudes, traditions, power relations, material constraints, and end-products together with the agency, artifice, and social relations of technicians' (Dobres 1999: 128).

How can we make better use of modern practice? This represents a concern for archaeology more than other disciplines because we are keenly interested in the production, use, and discard — in other words, the biography — of things in the past. At the heart of this issue is the relationship we have as academics with practical knowledge. Artefacts are at times studied with little understanding of how those objects came into being. By this I am referring not to provenance or structural composition but to fabrication in all its complexity: conceptualisation, design, and execution (Shanks & McGuire 1996). This underlines the problem of a paucity of experiential knowledge within the discipline. For a subject reliant on understanding how people made objects, we pay relatively little attention to those involved in contemporary manufacture and crafts. It is the exception rather than the rule to engage those skilled in the creation of objects, individuals who have spent years if not decades refining their understanding of a given production process.

I am not suggesting that practical know-how serves as a panacea for all the gaps in how we study production and products, particularly since there are pitfalls to avoid when requisitioning and utilising this knowledge. It is not enough to ask a contemporary crafts worker, 'Tell me what you know.' In much the same way that an academic lecture provides a distilled précis, a tranche, which needs to be supplemented with additional questions and debate, so too must we explore ways to derive practical knowledge. In this regard, the inclusion of anthropological methods of enquiry could be crucial, providing case studies and approaches that support why more attention should be paid to skilled production (Keller & Keller 1996; Marchand 2010). Production, as a step in the object's biography, can be more completely understood because it is subject to specific constraints. This then serves as a strong basis for interpreting use and discard, parameters that are subject to a wider variety of influences.

Actualistic approaches as they are currently undertaken – self contained and discrete, designed to address specific questions – are not a suitable substitute. Experiential knowledge built through experimental replications is an important, but essentially amateur, form of skill; this form of 'expertise' should not come to represent the totality of practical know-how in our subject. The cycle of building experimental knowledge on the basis of experimental replications is one that requires reconsideration and reevaluation, as it embodies a distorted view of practice that ultimately damages the discipline.

Developments in *chaîne opératoire* recognise the risks of neglecting 'practice' and steer us firmly in the direction of a better integration of knowledge and skill. How might archaeology alter with an influx of craft expertise, fostered from within the discipline, and bridging the gap between academic and craft expertise? One way to institutionalise broader practical training into the discipline would be through apprenticeships. Archaeological field schools are already a well-organised form of apprenticeship (Wallaert 2012: 6). Unfortunately, this type of learning experience does not form part of the pedagogy of archaeologists focused on material studies, even though there are many benefits to be had from learning to make and do.

Institutionalising apprenticeships in archaeology more generally is a radical suggestion, but not without precedent from within the discipline and in academia more generally. Walls investigated the relationship between the fabrication of kayaks and the skilled act of hunting, with its attendant social and economic ramifications. Of critical importance to his ability to interrogate the virtuosity of kayak construction was the fact that he was an experienced kayak builder in his own right (Walls 2016). Marchand undertook a two-year apprenticeship in woodworking as part of an investigation into the concept of 'embodied cognition', viewing and assessing the topic through the lens of skilled production (Marchand 2010). In turn, this was built on episodes of apprenticeship with minaret-builders in Yemen (Marchand 2001). Similarly, C. Keller undertook a six-month apprenticeship with a blacksmith – and continued to practice blacksmithing – in order to provide new insights and empirical tests into an 'anthropology of knowledge' (Keller & Keller 1996). In all these cases, integrating practical expertise has been more than a mere novelty; it has transformed the understanding of each topic in entirely innovative ways.

Keller and Keller's work exemplifies a long tradition. Apprenticeship has served as an anthropological training device since the 1970s. It was not uncommon for students to spend a significant portion of their graduate studies learning a craft that then provided the grounding for their thesis (Marchand 2008). However, this vocational experience has more or less vanished from the modern discipline, sounding a note of caution for archaeologists. How we view and valorise our own craft – field archaeology – has come under scrutiny for some time (Shanks & McGuire 1996; Berggren & Hodder 2003). The practical branches of our discipline, field training and experimental archaeology, have all too often been marginalised in favour of topics seen as more academic (Wendrich 2012: 6).

Integrating apprenticeship into our pedagogy will likely meet a stumbling block as archaeology students are increasingly in favour of learning lab-based skills. In response to these points, I cite an extremely important development in the discipline that has been largely overlooked. In part driven by the principles of *chaîne opératoire*, the whole field of lithic research has been revolutionised by one key advance: a few lithic analysts learnt to flint-knap (Johnson 1978; Bleed 2008).[1] The efforts of a relatively small number of pioneering archaeologists have made a radical difference to the way we conceive, study, and understand stone as a material for tool production and the tools as social objects.

For contemporary lithic studies, knowledge is built from within the subject through experimental knapping or from inference derived from the archaeological record. The opportunities for ethnographic research are limited considering the displacement of this technology from most societies around the world (see Knudson's in response to Johnson, in comments section of Johnson 1978). For butchery, we have a fundamentally different situation. The activity and the behaviour are still an integral part of the majority of modern cultures; we have a uniquely usable way to integrate fresh perspectives. Can we aspire to study the true complexity of butchery, as an ecosystem, without participating in the activity? Zooarchaeologists could benefit considerably by observing and participating in different forms of butchery as a way to better appreciate the mechanism through which bone assemblages are created.

This book has heavily emphasised the ways in which modern analogy has influenced the method I propose for recording butchery. More than this, it also shows how the contemporary context has been woven into the fabric of the approach. For example, in addition to the tools and techniques of carcass processing, an understanding of the biology of modern animals, as developed through both academic and practical training, has been used to literally 'flesh out' archaeological fauna. Makers keep learning during their entire lives (Wallaert 2012: 22), and in my case, one never stops being a butcher. Thus, above all, we need to valorise craft expertise before we are able to integrate the *connaissance* of craftspeople into our own frames of reference.

CONNECTING THE PAST TO THE PRESENT

The benefits of enhancing practical knowledge in archaeology extend beyond the discipline, offering a path to connect archaeological outcomes to contemporary cultures through anthropological enquiry (Walls 2016). The examples from Asia and Africa discussed in Chapter 4 signal a major gap. If entire subcultures can be constructed around slaughter and butchering, for that is what the cases illustrate, how can we have failed to see the full extent of the social context? The development and expression of outcaste status is only superficially related to the activity of those who form part of these groups. With one notable exception (a revealing account penned by a Mr Päk, head of the butchers' guild, representing a group of butchers from Seoul, Korea, and translated into English; see Moore 1894), my review has found scant attempts to portray the perspective of those involved in these trades. Very little attention has been devoted to the study of these groups, whether from the perspective of the history of the professions in which they are involved, the development of the craft, or the idiosyncrasies of a set of practices that must surely have become highly standardised. These are topics that archaeology and anthropology are ideally placed to investigate and which would prove fertile ground for developing a new understanding of the relationship between craft and society. Perhaps more important is the context of relevancy. The emphasis of current research has, rightly, been on the mistreatment and prejudice that these groups have endured (Kim 2013: 15–34; McCormack 2013: 30–59). A small measure of redress might be achieved through valorisation of what these groups do, rather than what is done to them.

At a more general level, illuminating the complexity of how butchers operate, how their skills influence social life, and how society views the butcher – the top-down view – provides a resounding reason for us to reconsider our own archaeological data – the bottom-up view. The types of cut marks observed at urban and military Roman sites (e.g., their patterning and the regularity) reveal specific details about 'the butcher' as a social persona and performer, in addition to the economic influence of butchery.

Delving more deeply into what constitutes a 'butcher', and the meaning of the butcher's craft, exemplifies the utility of looking well beyond the boundaries of archaeological research. Directly and indirectly, those involved in the processing of meat and carcass parts have been the topic of study from a range of perspectives. At least since *The Jungle* (Sinclair 1906), and especially from the 1960s onward, attention has been drawn to the conditions of work for those involved in the modern meat trade (Pacyga 2015: 63–93), the place of black and ethnic minorities in the US industry (Fogel 1970), and the reaction by those in the trade to their working conditions (Pilcher 2006). At the other end of the spectrum, in many parts of Europe and the United States, artisanal butchers are enjoying a resurgence of interest in their profession, seen as a more ethical and honest counter to industrialised practices (Jones 2015). These sources offer much that archaeologists can draw on to situate the social context within which butchers operated in the past.

Indeed, archaeology could become a living and growing repository for the knowledge of craftspeople, in a world that is increasingly succumbing to mechanisation. Butchery thus illustrates the utility of archaeology for a general audience. There is a desire to know how crafts developed (Jones 1976; Lee 2008; Perren 2008) and understand this aspect of our ancestors' relationship with animals. It is an idealised version of the craft, reimagined from the past, which is now enjoying a renaissance (Fearnley-Whittingstall 2007: 50–3; Ocejo 2014). The march for progress in industrialised societies has erased many aspects of our relationship with food procurement. In this case, it is not only mechanisation of food production but also a gamut of lifestyle choices that have marginalised our knowledge of 'crafting food'. We hunt and gather in supermarkets. In turn, our modern lifestyle has implications for our well-being and the relationship we have with food (Pollan 2008: 8–13).

Archaeological research describing the knowledge of practice offers opportunities to contribute to modern debates. By blending the social, technological, and practical, can we offer a deeper-time perspective that could help in addressing concerns regarding our contemporary consumption of meat?

As the discussion in Chapter 11 has shown, it has been possible to deduce which techniques of butchery were favoured, to outline the mode of disarticulation, and subsequently to provide an interpretation of the wider role of butchery during the periods in question. We observe a piecemeal pattern of disarticulation from the Romano-British assemblages, while quartering is noted on medieval sites. By refining our assessment of carcass division, we take an important step toward revealing the trajectory of transitions in economic drivers.

Using cases drawn from the Romano-British period: butchery at the scapula, including trimmed glenoid margins and perforated scapula blades, have been interpreted as evidence of curing the shoulder meat in either brine vats or smokers (Grant 1975; Dobney et al. 1999). Subsequent knicks to the surface of the blade suggest shaving cuts to remove slices of meat from the shoulder bone (Lauwerier 1988: 156). Evidence for extensive processing of bone for marrow has been observed from Gloucester (Levine 1986), Lincoln (Dobney et al. 1996), and York (O'Connor 1988; Carrott et al. 1995). The butchered assemblages from the sites I have studied indicate that the meat from most elements of the carcass, and in particular the long bones, was first filleted, and then the bones were systematically chopped. At Lincoln, this pattern occurred on a scale large enough to be considered 'truly commercial' (Dobney et al. 1999). This holistic view of 'Roman butchery' observes regional and diachronic transitions in unison, helping to shed light on the global phenomenon, revealing not only culinary preferences but also an attitude to meat and meat consumption. In this way, the archaeological record sets the stage for developments that culminate in contemporary practice.

By establishing precedent, we not only observe changes in our patterns of consumption and their origins, but we also have the opportunity to situate other transformations taking place around the activity. Intensified animal husbandry to meet increased meat consumption does not follow a linear relationship. Many influences are exerted, which include the organisation of production and dispersal; political implications on an economically significant industry; changing attitudes to animals and consumption of their flesh; changing social contexts around when, where, and how to consume; etc.

Archaeology helps to provide the larger picture. The questions archaeologists can pose are grounded in this more holistic framework. We have observed the diversification of the craft into specialist roles. Therefore, when we ask, 'What happens to carcass by-products today?' (a topic at the heart of waste control in the modern industry; Jayathilakan et al. 2012), the question is based on evidence of an increased marginalisation of specific parts of the animal, where once entire industries were based around some of these same body parts. In the same way, if we examine what mechanisation has meant for butchery and meat consumption, this is based on a long series of technological developments, culminating in robotic disassembly of the carcass (Purnell et al. 1990).

The main limiting factor for archaeologists is that the way cut marks have been studied to date leaves little room to connect with modern practice in meaningful ways, and even less scope to contribute to contemporary concerns. Food production has become the nexus around which issues of climate damage, human and animal health, and the ethical treatment of the animals revolve (Fitzgerald 2015). Meat is much debated because of the ethical and ecological ramifications of modern Western patterns of consumption, which have come to be identified with 'modernity' and, as such, increasingly adopted by newly industrialised nations (Jayathilakan et al. 2012).

In the United States alone, grain consumed by animals raised for meat could feed 840 million people (Pimentel 2004). This is a staggering figure that has ramifications for issues such as animal welfare, global nutritional imbalance, and climate change. Archaeology has the potential to contribute in a number of ways. An archaeological narrative could be critical for describing and explaining Western cosmology and dependence on the specific form of meat consumption engaged in today. This involves comparing and contrasting, both diachronically and synchronically, in order to better understand why meat has attained the significance it has. We may also be able to provide data to better understand the ecological context. The relationship between a warming climate consequential to our food production practices (de Boer et al. 2013) – 'heat and meat' – falls squarely within the purview of our discipline, unifying the strengths of environmental and interpretive archaeology. These points represent two significant avenues through which archaeologists can contribute to tackling problems of a universal nature.

However, a more fundamental way we can contribute is by exposing gaps in scholarly endeavour and transforming the conceptual basis from which researchers study the problems outlined above. To offer just one telling example, in an important recent analysis of the development of the feed industry, Coffey et al. overlook the importance of the deeper chronological context in a sweeping account of human–animal relations over millennia:

> Nobody knows when deliberate animal feeding systems developed, as it happened before the advent of writing. Techniques of animal husbandry developed spontaneously some 12,000 years ago ... Changes in agricultural production begun in the 18th century culminated in better approaches to overall feeding at around the turn of the 19th century.
>
> (Coffey et al. 2016: 1)

These comments generalise to the point of factual inaccuracy, giving the impression that researchers feel little need to contextualise important changes in human–animal relationships, transitions that are well described in the literature. Important archaeological research on the advent of domestication, the transition to agriculture, the specialisation of husbandry techniques, and the nuances of livestock improvement are all overlooked. The quote emphasises why the *longue durée* perspective is necessary, but based on the kind of fine-grained analysis advocated in this book, analysis that can lead to a diversity of revealing insights we would not otherwise have.

FINAL THOUGHTS

Butchery has influenced every culture, whether meat-eating or otherwise. The main thrust of this book has been to champion a deeper appreciation for the wider context of archaeological butchery in order to better assess human–animal relationships in the past. The growing call to embed better ways to study 'knowledge' and 'practice' in the discipline – whether as an aspect of craft production or to understand the people who made things – provides important opportunities not only to study the past but also to connect to the present. For butchery specifically, if we achieve a more holistic assessment of the drivers and outcomes of changes in meat processing, we create links to the wider historical and contemporary literature on meat, the meat trade, and our modern-day patterns of consumption. This path may ultimately lead to archaeologists having a more substantive voice that helps contextualise meat-eating today. However, as important as connections to the historic and anthropological contexts are, we have the potential to unite our data – indeed, it is incumbent that we do so – to the ecological and climatic research on these topics.

Finally, the book makes a resounding call for practicalism: for sustained attention, even devotion, to all things practical. This applies particularly to those of us who are material specialists and serves as both complement and antithesis to phenomenology. Unless we galvanise our efforts and enact change, archaeologists risk becoming academic spectators of practical knowledge. *If we become better at making, we become better at thinking about things made in the past.* This is the essential message I wish to convey in this book, woven throughout the descriptions and analyses of butchery contexts across time and location, and which I have exemplified through the experiential knowledge, describing personal skill and activity, that has imbued this book.

NOTES

Chapter 2

1 Animals use tools; the distinction here rests on the manner and complexity of use, and the agency driving the use of tools for butchery.

Chapter 4

1 Gandhi initiated a similar movement, employing and popularizing the word *Harijans*, meaning 'Children of God', which was subsequently seen as paternalistic and rejected, with *Dalit* – meaning 'crushed' – then adopted in its place (Natrajan 2012: 170n2).

2 The notion that 'caste' can be used to reference African contexts has been much debated; see Todd 1977.

Chapter 5

1 The accumulation of blood in the lower, or dependent, parts of the body. This is also known as *livor mortis*, from livor – 'bluish colour' – and mortis – 'of death' – due to the colour it causes.

2 Bone was an important raw material at least until World War II (Crabtree & Campana 2008: 326) and is still converted into bone meal and sold for feed, etc.; thus, it has economic value in the modern setting. We should also not forget regional preferences where offal has an elevated status, such as haggis in Scotland, or East European blood sausage, *kiszka*.

3 This discussion of aesthetics relates specifically to the modern Western world, and Britain in particular.

4 This raises the distinction between game and non-game animals. In the United Kingdom, for example, rabbit, like all game, is considered a more exclusive food. One also invariably purchases whole game birds, rather than portions.

5 These accounts are drawn from smallholder and mid-twentieth-century practical guides. The industrialisation of butchery served to undermine and marginalise these animal welfare concerns (Grandin 2010).

6 The meaning of this term is not immediately clear, although it seems to indicate meat from animals that have asphyxiated on their own blood (see Saunders 1990: 278–9 and n24). From personal experience, when an animal suffocates on its own blood following slaughter, the accumulated blood in the esophagus can form into a spongy solid as a consequence of clotting action. This may offer an alternative explanation, referencing the creation of this coagulated blood.

7 The difference between 'cut mark' and 'knife mark' will be discussed in more detail in Chapter 9.

Chapter 6

1 Some of the discussion is based in part on personal preference, i.e., those qualities that I favoured in knives during my period in the profession, as a way of explaining aspects of functionality. This also explains why I have not included any discussion of the axe, a tool that has been entirely superseded in the modern profession, and one with which I have little familiarity.

2 To clarify, in theory the knife is fully functional without a handle. However, in practice, the handle not only makes use of the knife more comfortable but allows for a more stable grip. These two factors facilitate butchery.

3 After 1914, steel alloys are not the same as carburised iron that we would find in archaeological contexts. Forms of modern steel, such as molybdenum, are much less likely to

corrode (stainless) and are generally stronger while being lighter (Tylecote 1992: 168).

4   A thick blade is desirable in a cleaver as it reduces the likelihood of chipping the edge; conversely, it is precisely these knicks in archaeological blades that leave characteristic striations that help the analyst to deduce directionality of the mark.

Chapter 7

1   The distinction is not well defined. Tools are studied from both metal and osteological materials, as the signatures of the tools have been an important indirect means of learning about the transition and adoption of metallurgy (Greenfield 2013a, 2016).

Chapter 8

1   Binford (1981) did use metal tools (but not assemblages created with metal implements) to construct his butchery recording protocols, but this was undertaken in a context that had little correlation with later period sites.

2   MacKinnon does synthesis the data on cut marks from sites in Italy; however, here I refer to patterning specifically, not proportions and occurrence.

3   Theodore White also used 'smashing' (White 1952: 338) as a way of describing the activity he recorded from Angosture Reservoir basin, South Dakota, an aboriginal site in the United States. This perhaps set a precedent for the type of descriptive terminology that places less emphasis on planned actions: to my mind the bones are not 'smashed' but deliberately fractured, using tools and precise force in order to recover or process a commodity, be it brains or marrow.

4   The distinction here is between a specialized implement created to chop, a cleaver, versus a general-purpose knife.

5   For small mammals and fowl, skillful use of a cleaver renders it serviceable for jointing and chopping, while leaving no bone splinters at all.

Chapter 9

1   I deliberately favour 'butchery' as opposed to cut marks, as this more accurately reflects the underlying driver, to record process.

2   There will be overlap in the descriptions that follow, as there is an obvious relationship between activity and tools. However, the point is that recording a chop mark is not the same as recording a cleaver mark. A chop can be made with a large blade, for example. The aim is to capture nuance, which then informs interpretation.

Chapter 10

1   Although these stages share many similarities with Rixson's earlier work (1988), they are structurally different in that they are specific to the pattern noted from Cirencester and are formulated based on activity categories that are not necessarily sequential.

Chapter 11

1   Variation need not preclude standardisation; for example, modern British butchery practices actually show a great deal of regional variability, although within each region there is a high degree of standardisation and accord in the butchery practices used.

2   Even in this case, the 'cut of meat' cannot be equated to the modern iteration. Guild practices may well have resulted in specific butchery techniques; however, we are unlikely to see these in the archaeological record.

3   Hence the very large commercial freezers to accommodate half- and quarter-carcasses – the meat could be cut into smaller units, but this then has implications for transportation, hygiene, meat quality, etc.

4   This practice is more pertinent for cattle than for other species owing to the amount of blood that remains in the carcass even after evisceration, which increases the probability of spoilage.

Chapter 12

1   One needs to be pragmatic about the realities of managing an academic career in the modern day, with increasing fees and a large proportion of students undertaking paid work. However, one mutually beneficial solution may in fact reside in undertaking an apprenticeship as paid work.

REFERENCES

Abe, Y., Marean, C. W., Nilssen, P., Stone, E., & Assefa, Z. 2002. The analysis of cutmarks on archaeofauna: a review and critique of quantification procedures, and a new image-analysis GIS approach. *American Antiquity* 67: 643–63.

Adams, C. 1990. *The Sexual Politics of Meat: A Feminist-Vegetarian Critical Theory*. New York, NY, Continuum.

Adams, J. 2010. *Gandhi: Naked Ambition*. London, Quercus.

Aeillo, L., & Wheeler, P. 1995. The expensive-tissue hypothesis: the brain and the digestive system in human and primate evolution. *Current Anthropology* 36(2): 199–221.

Ahmed, Z. 2013. Marginal occupations and modernizing cities: Muslim butchers in urban India. *Economic and Political Weekly* 48(32): 121–31.

Aird, P. M. 1985. On distinguishing butchery from other post-mortem destruction: a methodological experiment applied to a faunal sample from Roman Lincoln. In Fieller, N. R., Gilbertson, D. D., and Ralph, N. G. (eds.), *Palaeobiological Investigation*. Oxford, BAR Publishing: 5–35.

Albarella, U. 1997. Shape variation of cattle metapodials: age, sex or breed? Some examples from mediaeval and postmediaeval sites. *Anthropozoologica* 25–6: 37–47.

Albarella, U. 2003. Tanners, tawyers, horn working and the mystery of the missing goat. In Murphy, P., & Wiltshire, P. (eds.), *The Environmental Archaeology of Industry*. Oxford, Oxbow Books: 71–86.

Albarella, U., Johnstone, C., & Vickers, K. 2008. The development of animal husbandry from the Late Iron Age to the end of the Roman period: a case study from South-East Britain. *Journal of Archaeological Science* 35(7): 1828–48.

Alcock, J. P., & Pilsbury, J. 1996. *Book of Life in Roman Britain*. London, B. T. Batsford.

Ambrose, S. H. 2001. Paleolithic technology and human evolution. *Science* 291(5509): 1748–53.

Amos, T. D. 2011. *Embodying Difference: The Making of Burakumin in Modern Japan*. Honolulu, HI, University of Hawai'i Press.

Andrefsky, W. 2005. *Lithics*. Cambridge, Cambridge University Press.

Appadurai, A. 1988. Putting hierarchy in its place. *Cultural Anthropology* 3(1): 36–49.

Arbuckle, B. S., & McCarty, S. A. 2014. *Animals and Inequality in the Ancient World*. Boulder, CO, University Press of Colorado.

Århem, K. 1987. *Milk, Meat and Blood: Diet as a Cultural Code among the Pastoral Maasai*. Uppsala, Working Papers in African Studies.

Arnold, D. E. 1975. Ceramic ecology of the Ayacucho Basin, Peru: implications for prehistory. *Current Anthropology* 16(2): 183–205.

Arthur, J. W. 2014a. Pottery uniformity in a stratified society: an ethnoarchaeological perspective from the Gamo of southwest Ethiopia. *Journal of Anthropological Archaeology* 35: 106–16.

Arthur, J. W. 2014b. Culinary crafts and foods in southwestern Ethiopia: an ethnoarchaeological study of Gamo groundstones and pottery. *African Archaeological Review* 31(2): 131–68.

Arthur, K. W. 2008. The Gamo hideworkers of southwestern Ethiopia and cross-cultural comparisons. *Anthropozoologica* 43(1): 67–98.

Arthur, K. W., Arthur, J. W., Curtis, M. C., Lakew, B., & Lesur-Gebremarium, J. 2009. Historical archaeology in the highlands of southern Ethiopia: preliminary findings. *Nyame Akuma* 72: 3–11.

Arzt, J., White, W. R., Thomsen, B. V., & Brown, C. C. 2010. Agricultural diseases on the move early in the third millennium. *Veterinary Pathology* 47(1): 15–27.

Ashbrook, F. G. 1955. *Butchering, Processing and Preservation of Meat*. New York, NY, Van Nostrand Reinhold.

Astill, G. 1998. Medieval and later: composing an agenda. In Bayley, J. (ed.), *Science in Archaeology: An Agenda for the Future*. London, English Heritage: 168–79.

Atici, A. L. 2006. Middle-range theory in Paleolithic archaeology: the past and the present. *Journal of Taphonomy* 4(1): 29–45.

Audoin-Rouzeau, F. 1987. Medieval and early modern butchery: evidence from the monastery of La Charite-Sur-Loire (Nievre). *Food and Foodways* 2: 31–48.

Audouze, F. 2002. Leroi-Gourhan, a philosopher of technique and evolution. *Journal of Archaeological Research* 10: 277–306.

Avent, R. 1979. Laugharne Castle 1978. *The Carmarthenshire Antiquary* 15: 39–56.

Avent, R. 1995. *Laugharne Castle*. Cardiff, CADW, Welsh Historic Monuments.

Avent, R., & Read, E. 1977. Laugharne Castle 1976. *The Carmarthenshire Antiquary* 13: 17–41.

Baillie-Grohman, A. F. 1904. *The Master of the Game*. London, Chatto & Windus.

Baker, P., & Worley, F. (eds.) 2014. *Animal Bones and Archaeology: Guidelines for Best Practice*. Portsmouth, English Heritage.

Barber, G. 1999. The animal bones. In Davenport, P. (ed.), *Archaeology in Bath Excavations 1984–89*. Oxford, BAR Publishing: 107–15.

Bar-Oz, G., & Munro, N. D. 2007. Gazelle bone marrow yields and Epipalaeolithic carcass exploitation strategies in the southern Levant. *Journal of Archaeological Science* 34(6): 946–56.

Barthes, R. 1975. Towards a psychosociology of contemporary food consumption. In Forster, E., & Forster, R. (eds.), *European Diet: From Pre-Industrial to Modern Times*. New York, Harper & Row: 47–59.

Barton, H., Barker, G., Gilbertson, D., Hunt, C., Kealhofer, L., Lewis, H., Paz, V., Piper, P.,

Rabett, R., Reynolds, T., & Szabo, K. 2013. Late Pleistocene foragers, c. 35,000–11,500 years ago. In Barker, G. (ed.), *Rainforest Foraging and Farming in Island Southeast Asia: The Archaeology of the Niah Caves, Sarawak*. Cambridge, McDonald Institute for Archaeological Research: vol. 1: 171–215.

Bartosiewicz, L. 2003. 'There's something rotten in the state . . .': bad smells in antiquity. *European Journal of Archaeology* 6(2): 175–95.

Bartosiewicz, L. 2008a. Taphonomy and palaeopathology in archaeozoology. *Geobios* 41(1): 69–77.

Bartosiewicz, L. 2008b. Description, diagnosis and the use of published data in animal palaeopathology: a case study using fractures. *Veterinarija ir Zootechnika* 41(63): 12–23.

Bartosiewicz, L. 2009. Skin and bones: taphonomy of a medieval tannery in Hungary. *Journal of Taphonomy* 7(2–3): 91–107.

Bartosiewicz, L., Neer, W. V., & Lentacker, A. 1997. *Draught Cattle: Their Osteological Identification and History*. Tervuren, Musée Royal de L'Afrique Centrale.

Bar-Yosef, O., & Van Peer, P. 2009. The chaîne opératoire approach in Middle Paleolithic archaeology. *Current Anthropology* 50(1): 103–31.

Bealer, A. 1972. *Old Ways of Working Wood*. Barre, MA, Barre.

Belcastro, G., Rastelli, E., Mariotti, V., Consiglio, C., Facchini, F., & Bonfiglioli, B. 2007. Continuity or discontinuity of the life-style in central Italy during the Roman Imperial Age–Early Middle Ages transition: diet, health, and behavior. *American Journal of Physical Anthropology* 132: 381–94.

Bell, J. A. 1994. *Reconstructing Prehistory*. Philadelphia, PA, Temple University Press.

Bello, S. M. 2011. New results from the examination of cut-marks using three-dimensional imaging. In Ashton, N., Lewis, S. G., & Stringer, C. (eds.), *The Ancient Human Occupation of Britain*. Amsterdam, Elsevier: 249–62.

Bello, S. M., & Soligo, C. 2008. A new method for the quantitative analysis of cutmark micromorphology. *Journal of Archaeological Science* 35(6): 1542–52.

Benco, N. L., Ettahiri, A., & Loyet, M. 2002. Worked bone tools: linking metal artisans and animal processors in medieval Islamic Morocco. *Antiquity* 76: 447–57.

Berggren, Å., & Hodder, I. 2003. Social practice, method, and some problems of field archaeology. *American Antiquity* 68(3): 421–34.

Bettinger, R. L., Garvey, R., & Tushingham, S. 2015. *Hunter-Gatherers: Archaeological and Evolutionary Theory.* New York, NY, Springer.

Billington, S. 1990. Butchers and fishmongers: their historical contributions to London's festivity. *Folklore* 10: 97–104.

Binford, L. R. 1977. *For Theory Building in Archaeology: Essays on Faunal Remains, Aquatic Resources, Spatial Analysis, and Systemic Modeling.* New York, NY, Academic Press.

Binford, L. R. 1978. *Nunamiut Ethnoarchaeology.* London, Academic Press.

Binford, L. R. 1981. *Bones: Ancient Men and Modern Myths.* New York, NY, Academic Press.

Binford, L. R. 1983. *Working at Archaeology.* New York, NY, Academic Press.

Binford, L. R. 1984. Butchering, sharing, and the archaeological record. *Journal of Anthropological Archaeology* 3(3), 235–57.

Binford, L. R. 2001. *Constructing Frames of Reference: An Analytical Method for Archaeological Theory Building Using Hunter-Gatherer and Environmental Data Sets.* Berkeley, CA, University of California Press.

Birley, R. 1977. *Vindolanda: A Roman Frontier Post on Hadrian's Wall.* London, Thames and Hudson.

Blakelock, E., & McDonell, J. G. 2007. A review of metallographic analyses of early medieval knives. *Historical Metallurgy* 41(1): 40–56.

Blech, Z. Y. 2009. *Kosher Food Production.* Ames, IA, John Wiley & Sons.

Bleed, P. 1991. Operations research and archaeology. *American Antiquity* 56(1): 19–35.

Bleed, P. 2008. Skill matters. *Journal of Archaeological Method and Theory* 15(1): 154–66.

Blumenschine, R. J., Marean, C. W., & Capaldo, S. D. 1996. Blind tests of interanalyst correspondence and accuracy in identification of cut marks, percussion marks and carnivore tooth marks on bone surfaces. *Journal of Archaeological Science* 23: 493–507.

Bond, J. M., & O'Connor, T. P. 1999. *Bones from Medieval Deposits at 16–22 Coppergate and Other Sites in York.* York, Council for British Archaeology.

Boon, G. C. 1957. *Roman Silchester: The Archaeology of a Romano-British Town.* London, Max Parrish.

Bouchnik, R. 2016. Meat consumption patterns as an ethnic marker in the late second century: comparing the Jerusalem City dump and Qumran assemblages. In Marom, N., Yeshuran, R., Weissbrod, L., & Bar-Oz, G. (eds.), *Bones and Identity: Zooarchaeological Approaches to Reconstructing Social and Cultural Landscapes in Southwest Asia.* Oxford, Oxbow: 303–23.

Bourdieu, P. 1990. *The Logic of Practice.* Stanford, CA, Stanford University Press.

Brain, C. K. 1981. *The Hunters or the Hunted?* Chicago, IL, University of Chicago Press.

Brantz, D. 2005. Animal bodies, human health, and the reform of the slaughterhouses in nineteenth-century Berlin. *Food & History* 3(2): 193–215.

Braun, D., R. Pobiner, B. L., & Thompson, J. C. 2008. An experimental investigation of cut mark production and stone tool attrition. *Journal of Archaeological Science* 35: 1216–23.

Brenet, M., Chadelle, J.-P., Claud, E., Colonge, D., Delagnes, A., Deschamps, M., Folgado, M., Gravina, B., & Ihuel, E. 2017. The function and role of bifaces in the late middle Paleolithic of southwestern France: examples from the Charente and Dordogne to the Basque country. *Quaternary International* 428: 151–69.

Bridbury, A. R. 2008. *Medieval England: A Survey of Social and Economic Origins and Development.* Leicester, Troubador.

Brink, J. W. 1997. Fat content in leg bones of *Bison bison*, and applications to archaeology. *Journal of Archaeological Science* 24: 259–74.

Brisebarre, A. M. 1998. *La Fête du Mouton: Un Sacrifice Musulman dans l'Espace Urbain.* Paris, CNRS Éditions, 351.

Bromage, T. G., & Boyde, A. 1984. Microscopic criteria for the determination of directionality of cutmarks on bone. *American Journal of Physical Anthropology* 65(4): 359–66.

Brumberg-Kraus, J. 1999. Meat-eating and Jewish identity: ritualization of the priestly 'Torah of Beast and Fowl' (Lev 11:46) in Rabbinic Judaism and medieval Kabbalah. *Association of Jewish Studies Review* 24(2): 227–62.

Buckser, A. 1999. Eating and social identity among the Jews of Denmark. *Ethnology* 38(3): 191–209.

Bunn, H. 1981. Archaeological evidence for meat-eating: by Plio-Pleistocene hominids from Koobi Fora and Olduvai Gorge. *Nature* 291: 574–77.

Bunn, H., Bartrum, L., & Kroll, E. 1988. Variability on bone assemblage formation from Hadza hunting, scavenging, and carcass processing. *Journal of Anthropological Archaeology* 7: 412–57.

Bunn, H., & Kroll, E. 1986. Systematic butchery by Plio/Pleistocene hominids at Olduvai Gorge, Tanzania. *Current Anthropology* 27(5): 431–52.

Callon, M. 1984. Some elements of a sociology of translation: domestication of the scallops and the fishermen of St. Brieuc Bay. *The Sociological Review* 32(S1): 196–233.

Campana, D. V., & Crabtree, P. J. 2014. Worked bone objects from the Iron Age site of Kyzyltepa, Uzbekistan. In Xiaolin, M., & Yanfeng, H. (eds.), *Proceedings of the 9th Meeting of the (ICAZ) Worked Bone Research Group, Zhengzhou, China, 2013*. Beijing, Cultural Relics Press: 57–63.

Carlin, M., & Rosenthall, J. T. 1998. *Food and Eating in Medieval Europe*. London, Hambledon Press.

Carpenter, D. 2004. *The Struggle for Mastery: The Penguin History of Britain 1066–1284*. London, Penguin.

Carrott, J., Dobney, K. M., Hall, A., Issitt, M., Jaques, D., Johnstone, C., Kenward, H., Large, F., Mckenna, B., & Miles, A. 1995. *Assessment of Biological Remains from Excavations at Wellington Row, York (YAT/York Museum Code 1988–9.24)*. York, Reports from the Environmental Archaeology Unit.

Carter, E. H., & Mears, R. A. F. 2010. *A History of Britain*. London, Stacy International.

Cartmill, M. 1993. *A View to Death in the Morning: Hunting and Nature through History*. Cambridge, MA, Harvard University Press.

Carver, M. 2008. *Portmahomack: Monastery of the Picts*. Edinburgh, Edinburgh University Press.

Cessford, C. Alexander, M., & Dickens, A. 2006. *Between Broad Street and the Great Ouse: Waterfront Archaeology in Ely*. Cambridge, Cambridge Archaeological Unit.

Chadwick, E. 1843. *Report on the sanitary conditions of the labouring population of Great Britain: A supplementary report on the results of a special inquiry into the practice of interment in towns. Made at the request of Her Majesty's principal secretary of state for the Home department.* London: Printed by W. Clowes and Sons for H. M. Stationery Office.

Chapman, R., 2003. *Archaeologies of Complexity*. London, Routledge.

Chaiklin, S., & Lave, J. 1993. *Understanding Practice: Perspectives on Activity and Context*. Cambridge, Cambridge University Press.

Chaix, L., Dubosson, J., & Honegger, M. 2012. Bucrania from the Eastern Cemetery at Kerma (Sudan) and the practice of cattle horn deformation. In Kabaciński, J., Marek Chłodnicki, M., & Kobusiewicz, M. (eds.), *Prehistory of Northeastern Africa, New Ideas and Discoveries*. Poznan, Poznan Archaeological Museum: 185–208.

Chakravarti, M. 1979. Beef-eating in ancient India. *Social Scientist* 7(11): 51–5.

Charsley, S. 1996. 'Untouchable': what is in a name? *Journal of the Royal Anthropological Institute* 2(1): 1–23.

Chaudieu, G. 1966. *Boucher Qui Es-Tu? Ou Vas-Tu? On La Fabuleuse Histore des Bouchers, Celle d'Hier, d'Aujourd'hui et de Demain, Suivie du Memorial*. Paris, Peyronnet.

Chazan, M. 2009. Pattern and technology: why the chaîne opératoire matters. In Shea, J. J., & Lieberman, D. E. (eds.), *Transitions in Prehistory: Essays in Honor of Ofer Bar-Yosef*. Oxford, Oxbow: 469–78.

Choyke, Alice M. 2013. Hidden agendas: ancient raw material choice for worked osseous objects in Central Europe and beyond. In Choyke, A. M., & O'Connor, S. (eds.), *From These Bare Bones: Raw Materials and the Study of Worked Osseous Objects*. Oxford, Oxbow: 1–11.

Choyke, A. M., & O'Connor, S. (eds). 2013. *From These Bare Bones: Raw Materials and the Study of Worked Osseous Objects*. Oxford, Oxbow.

Choyke, A. M., & Schibler, J. 2007. Prehistoric bone tools and the archaeozoological perspective: research in Central Europe. In St-Pierre, C. G., & Walker, R. B. (eds.), *Bones as Tools: Current Methods and Interpretations in Worked Bone Studies*. Oxford, Archaeopress: 51–65.

Cioranescu. A. 1977. *Historia de la Conquista de las Siete Islas de Canarias*. Tenerife, Goya Ediciones.

Clark, J. G. D. 1952. *Prehistoric Europe: The Economic Basis*. London, Methuen.

Cleere, H. 1976. Iron making. In Strong, D., & Brown, D. (eds.), *Roman Crafts*. London, Duckworth: 127–53.

Clutton-Brock, J., & Armitage, P. L. 1977. Mammal remains from Trench A. In Blurton, T. R. (ed.), *Excavations at Angel Court, Walbrook, 1974*. London, Transactions of the London and Middlesex Archaeological Society: 14–100.

Coffey, D., Dawson, K., Ferket, P., & Connolly, A. 2016. Review of the feed industry from a historical perspective and implications for its future. *Journal of Applied Animal Nutrition* 4 (e3): 1–11.

Cohn-Sherbok, D. 2006. Hope for the animal kingdom: a Jewish vision. In Waldau, P., & Patton K. (eds.), *A Communion of Subjects: Animals in Religion, Science and Ethics*. New York, NY, Columbia University Press: 81–91.

Coles, J. 1979. *Experimental Archaeology*. London, Academic Press.

Coolidge, F. L., Wynn, T., Overmann, K. A., & Hicks, J. M. 2015. Cognitive archaeology and the cognitive sciences. In Bruner, E. (ed.), *Human Paleoneurology*. New York, NY, Springer International: 177–208.

Collingwood, R. G., & Richmond, I. 1969. *The Archaeology of Roman Britain*. London, Methuen.

Cope, C. 2004. The butchering patterns of Gamla and Yodefat: beginning the search for *kosher* practices. In O'Day, S., Neer, W. V., & Ervynck, A. (eds.), *Behaviour behind Bones: The Zooarchaeology of Ritual, Religion, Status and Identity*. Oxford, Oxbow Books: 25–34.

Corbin, A. 1995. The blood of Paris: reflections on the geneaology of the image of the capital. In Corbin, A. (ed.), *Times, Desire and Horror*. Cambridge, Cambridge University Press: 170–82.

Corsair, B. A., & Fitzell, W. L. 1975. *The York Butchers' Gild*. York, Ebor Press.

Cowgill, J., Neergaard, M. D., & Griffiths, N. 2001. *Knives and Scabbards*. London: Museum of London.

Coy, J., & Maltby, J. M. 1984. Archaeozoology in Wessex. In Keeley, H. C. M. (ed.), *Environmental Archaeology: A Regional Review*. London, English Heritage: 204–51.

Crabtree, P. 1989. *West Stow, Suffolk: Early Anglo-Saxon Animal Husbandry*. Ipswich, East Anglian Archaeological Reports.

Crabtree, P. J., & Campana, D. V. 2008. Traces of butchery and bone working. In Adams, B. J., & Crabtree, P. J. *Comparative Skeletal Anatomy* (pp. 323–45). Totowa, NJ, Humana Press.

Crabtree, P. J., Campana, D. V., & Ryan, K. 1989. *Early Animal Domestication and Its Cultural Context* (vol. 6). Philadelphia, PA, University of Pennsylvania Museum of Archaeology.

Crane, S. 2013. *Animal Encounters: Contacts and Concepts in Medieval Britain*. Philadelphia, PA, University of Pennsylvania Press.

Creighton, J. 2001. The Iron Age–Roman transition. In James, S., & Millett, M. (eds.), *Britons and Romans: Advancing an Archaeological Agenda*. York, CBA Research Report, vol. 125: 4–12.

Daly, P. 1969. Approaches to faunal analysis in archaeology. *American Antiquity* 34(2): 146–53.

Daniel, G. 1963. *The Idea of Prehistory*. Cleveland, OH, World Publishing.

Das, N. 2010. *Ambedkar, Gandhi and the Empowerment of the Dalit*. Jaipur, ABD Publishers.

Darvill, T. C., & Gerrard, C. M. 1994. *Cirencester: Town and Landscape*. Cirencester, Cotswold Archaeological Trust.

Daubney, R., Hudson, J. R., & Garnham, P. C. 1931. Enzootic hepatitis or Rift Valley fever: an undescribed virus disease of sheep cattle and man from East Africa. *Journal of Pathology and Bacteriology* 34(4): 545–79.

David, N., & Kramer, C. 2001. *Ethnoarchaeology in Action*. Cambridge, Cambridge University Press.

David, S. 2008. Zooarchaeological evidence for Moslem and Christian improvements of sheep and cattle in Portugal. *Journal of Archaeological Science* 35: 991–1010.

Davies, F. G. 2010. The historical and recent impact of Rift Valley fever in Africa. *American Journal of Tropical Medicine and Hygiene* 83(2 Suppl.): 73–4.

Day, J. N. 2005. Butchers, tanners, and tallow chandlers: the geography of slaughtering in early nineteenth-century New York City. *Food and History* 3(2): 81–102.

De Backer, C. J., & Hudders, L. 2015. Meat morals: relationship between meat consumption consumer attitudes towards human and animal welfare and moral behavior. *Meat Science* 99: 68–74.

de Boer, J., Schösler, H., & Boersema, J. J. 2013. Climate change and meat eating: an inconvenient couple? *Journal of Environmental Psychology* 33: 1–8.

De-Cupere, B., Lentacker, A., Neer, W. V., Waelkens, M., & Verslype, L. 2000. Osteological evidence for the draught use of cattle: first application of a new methodology. *International Journal of Osteoarchaeology* 10: 254–67.

de Garine, I. 2004. The trouble with meat: an ambiguous food. *Estudios del Hombre* 19: 33–54.

Derevenski, J. S. 2000. Rings of life: the role of early metalwork in mediating the gendered life course. *World Archaeology* 31(3): 389–406.

De Riaz, V. 1978. *The Book of Knives*. New York, NY, Crown.

DeVore, I., & Tooby, J. 1987. The reconstruction of hominid behavioral evolution through strategic modeling. In Kinzey, W. G. (ed.), *Evolution of Human Behavior: Primate Models*. Buffalo, NY, SUNY Press: 183–237.

De Vos, G. A., & Wagatsuma, H. 1966. *Japan's Invisible Race: Caste in Culture and Personality*. Berkeley, CA, University of California Press.

Dewbury, A., & Russell, N. 2007. Relative frequency of butchering cutmarks produced by obsidian and flint: an experimental approach. *Journal of Archaeological Science* 34: 354–7.

Diamond, J. 2002. Evolution, consequences and future of plant and animal domestication. *Nature* 418(6898): 700–7.

Dobney, K. 2001. A place at the table: the role of vertebrate zooarchaeology within a Roman research agenda for Britain. In James, S., & Millett, M. (eds.), *Britons and Romans: Advancing an Archaeological Agenda*. York, CBA Research Report, vol. 125: 36–46.

Dobney, K., Jaques, D., & Irving, B. 1996. *Of Butchers and Breeds: Report on Vertebrate Remains from Various Sites in the City of Lincoln*. Lincoln, City of Lincoln Archaeology Unit.

Dobney. K., Kenward, H., Ottaway, P., & Donel, L. 1998. Down, but not out: biological evidence for complex economic organization in Lincoln in the late 4th century. *Antiquity* 72: 417–24.

Dobney, K., Hall, A., & Kenward, H. 1999. It's all garbage: a review of bioarchaeology in the four English colonial towns. *Journal of Roman Archaeology Supplementary Series* 36: 15–36.

Dobres, M. A. 1999. Technology's links and *chaînes*: the processual unfolding of technique and technician. In Dobres, M. A., & Homan, C. R. (eds.), *The Social Dynamics of Technology: Practice, Politics and World Views*. Washington, DC, Smithsonian Institute.

Dobres, M. 2000. *Technology and Social Agency: A Framework for Archaeology*. Oxford, Blackwell.

Dominguez-Rodrigo, M. 1997. Meat-eating by early hominids at the FLK 22 Zinjanthropus site, Olduvai Gorge (Tanzania): an experimental approach using cut-mark data. *Journal of Human Evolution* 33(6): 669–90.

Domínguez-Rodrigo, M. 2002. Hunting and scavenging by early humans: the state of the debate. *Journal of World Prehistory* 16(1): 1–54.

Domínguez-Rodrigo, M. 2008. Conceptual premises in experimental design and their bearing on the use of analogy: an example from experiments on cut marks. *World Archaeology* 40(1): 67–82.

Domínguez-Rodrigo, M., Pickering, T. R., Semaw, S., & Rogers, M. J. 2005. Cutmarked bones from Pliocene archaeological sites at Gona, Afar, Ethiopia: implications for the function of the world's oldest stone tools. *Journal of Human Evolution* 48(2): 109–21.

Domínguez-Rodrigo, M., Pickering, T. R., & Bunn, H. T. 2012. Experimental study of cut marks made with rocks unmodified by human flaking and its bearing on claims of ∼3.4-million-year-old butchery evidence from Dikika, Ethiopia. *Journal of Archaeological Science* 39(2): 205–14.

Done, G. 1986. The animal bones from areas A & B. In Millet, M., & Graham, D. (eds.), *Excavations on the Romano-British small town at Neatham Hampshire, 1969–79*. Hampshire, Hampshire Field Club in Co-operation with the Farnham and District Museum Society: 141–7.

Donoghue, J. D. 1957. An Eta community in Japan: the social persistence of outcaste groups. *American Anthropologist* 59(6): 1000–17.

Dorrington, E. J. 1998. Bone report. In Clark, F. R. (ed.), *The Romano-British Settlement at Little London, Chigwell*. Essex, West Essex Archaeological Group: 109–12.

Douglas, M. 1970. *Natural Symbols*. London, Barrie & Rockliff.

Douglas, M. 1975. *Implicit Meanings*. London, Routledge & Kegan Paul.

Dumont, L. 1980. *Homo Hierarchicus: The Caste System and Its Implications*. Chicago, IL, University of Chicago Press.

Eckardt, H. 2014. *Objects and Identities: Roman Britain and the North-Western Provinces*. Oxford, Oxford University Press.

Edwards, J. 1984. *The Roman Cookery of Apicius*. London, Rider.

Efremov, I. A. 1940. Taphonomy: a new branch of paleontology. *Pan-American Geologist* 74(2): 81–93.

Elder, J. W. 1966. Fatalism in India: a comparison between Hindus and Muslims. *Anthropological Quarterly* 39(3): 227–43.

Ellen, R. 1994. Modes of subsistence: hunting and gathering to agriculture and pastoralism. In Ingold, T. (ed.), *Companion Encyclopedia of Anthropology, Humanity, Culture and Social Life*. London, Routledge: 197–226.

Elmasry, G., Barbin, D. F., Sun, D. W., & Allen, P. 2012. Meat quality evaluation by hyperspectral imaging technique: an overview. *Critical Reviews in Food Science and Nutrition* 52(8): 689–711.

Ervynck, A. 2004. Offrant, pugnant, laborant: the diet of the three orders in the feudal society of medieval north-western Europe. In O'Day, S. J., Van Neer, W., & Ervynck, A. (eds.), *Behaviour behind Bones: The Zooarchaeology of Ritual, Religion, Status and Identity*. Oxford, Oxbow: 215–23.

Fairlie, S. 2010. *Meat: A Benign Extravagance*. Hampshire, Permanent Publications.

FAO. 1991. *Guidelines for Slaughter, Meat Cutting and Further Processing*. Rome, Food and Agriculture Organization.

Fearnley-Whittingstall, H. 2007. *River Cottage Meat Book*. Berkeley, CA, Ten Speed Press.

Feeley-Harnik, G. 1995. Religion and good: an anthropological perspective. *Journal of the American Academy of Religion* 63(3): 565–82.

Ferguson, P. P., & Zukin, S. 1995. What's cooking? *Theory and Society* 24(2): 193–9.

Fiddes, N. 1991. *Meat: A Natural Symbol*. London, Routledge.

Fiona, A., & Mackenzie, D. 2000. Contested ground: colonial narratives and the Kenyan environment, 1920–1945. *Journal of Southern African Studies* 26(4): 697–718.

Fisher, J. W. 1995. Bone surface modifications in zooarchaeology. *Journal of Archaeological Method and Theory* 2(1): 7–68.

Fishkoff, S. 2010. *Kosher Nation: Why More and More of America's Food Answers to a Higher Authority*. New York, NY, Schocken.

Fitzgerald, A. J. 2010. A social history of the slaughterhouse: from inception to contemporary implications. *Human Ecology Review* 17(1): 58–69.

Fitzgerald, A. J. 2015. *Animals as Food: (Re)Connecting Production, Processing, Consumption, and Impacts*. East Lansing, MI, Michigan State University Press.

Flannery, K., & Marcus, J. 1998. Cognitive archaeology. In Whitley, D. S. (ed.), *Reader* in *Archaeological Theory: Post-Processual and Cognitive Approaches*. New York, NY, Routledge: 35–49.

Flower, B., & Rosenbaum, E. 1978. *The Roman Cookery Book*. London, Harrap.

Fogel, W. A. 1970. *The Negro in the Meat Industry*. Philadelphia, PA, University of Pennsylvania Press.

Forshaw, A., & Bergström, T. 1980. *Smithfield: Past and Present*. London, Heinemann.

Forth, G. 2007. Can animals break taboos?: Applications of 'taboo' among the Nage of eastern Indonesia. *Oceania* 77(2): 215–31.

Frachetti, M. 2002. Bronze Age exploitation and political dynamics of the eastern Eurasian steppe zone. In Boyle, K., Renfrew, C., & Levine, M. (eds.), *Ancient Interaction: East and West in Eurasia*. Cambridge, McDonald Institute for Archaeological Research: 161–70.

Frayn, J. 1995. The Roman meat trade. In Wilkins, J., Harvey, D., & Dobson, M. (eds.), *Food in Antiquity*. Exeter, University of Exeter Press: 107–14.

Frison, G. C. 1989. Experimental use of Clovis weaponry and tools on African elephants. *American Antiquity* 54(4): 766–84.

Fulford, M. 1991. Britain and the Roman Empire: the evidence for regional and long distance trade. In Jones, R. F. J. (ed.), *Roman Britain: Recent Trends*. Sheffield, J. R. Collis: 35–47.

Fuseini, A., Knowles, T. G., Hadley, P. J., & Wotton, S. B. 2016. Halal stunning and slaughter: criteria for the assessment of dead animals. *Meat Science* 119: 132–7.

Gamble, C. 1998. Palaeolithic society and the release from proximity: a network approach to intimate relations. *World Archaeology* 29(3): 426–49.

Garnier, T. 1921. *Les Grands Travaux de la Ville de Lyon: Études, Projets et Travaux Exécutés (Hôpitaux, Écoles, Postes, Abattoirs, Habitations en Commun, Stade, etc.)* Paris, Massin.

Geneste, J. M. 1990. Développement des systèmes de production lithique au cours du Paléolithique moyen en Aquitaine septentrionale. In *Paléolithique Moyen Récent et Paléolithique Supérieur Ancien en Europe*. Paris,

Mémoires du Musée de Préhistoire d'Ile-de-France 3: 203–13.

Gerrard, F. 1955. The craft of the butcher. In Gerrard, F. (ed.), *The Book of the Meat Trade*. London, Caxton: 1–3.

Gerrard, F. 1964. *Meat Technology*, 2nd edn. London, Leonard Hill.

Gerrard, F. 1977. *Meat Technology*, 3rd edn. London, Leonard Hill.

Gerrard, F. 1979. *Meat Technology*. Hove, Wayland.

Gidney, L. 2000. Economic trends, craft specialisation and social status: bone assemblages from Leicester. In Rowley-Conwy, P. (ed.), *Animal Bones, Human Societies*. Oxford, Oxbow Books: 170–9.

Gifford, D. P. 1981. Taphonomy and paleoecology: a critical review of archaeology's sister disciplines. *Advances in Archaeological Method and Theory* 4: 365–438.

Gifford, D. P., & Crader, D. C. 1977. A computer coding system for archaeological faunal remains. *American Antiquity* 42(2): 225–38.

Gifford-Gonzalez, D. 1991. Bones are not enough: analogues, knowledge, and interpretive strategies in zooarchaeology. *Journal of Anthropological Archaeology* 10(3): 215–54.

Gifford-Gonzalez, D. 1993. Gaps in zooarchaeological analysis of butchery: is gender an issue? In Hudson, J. (ed.), *From Bones to Behavior: Ethnoarchaeological and Experimental Contributions to the Interpretation of Faunal Remains*. Carbondale, Southern Illinois University Press: 181–200.

Gillingham, J., & Griffiths, R. A. 2000. *Medieval Britain: A Very Short Introduction*. Oxford, Oxford University Press.

Goldberg, P. J. P. 1992. *Women, Work and Life Cycle in a Medieval Economy*. Oxford, Clarendon Press.

González, M. Á. M., Yravedra, J., González-Aguilera, D., Palomeque-González, J. F., & Domínguez-Rodrigo, M. 2015. Micro-photogrammetric characterization of cut marks on bones. *Journal of Archaeological Science* 62: 128–42.

Goody, J. 1982. *Cooking, Cuisine and Class*. Cambridge, Cambridge University Press.

Gosselain, O. P. 2000. Materializing identities: an African perspective. *Journal of Archaeological Method and Theory* 7(3): 187–217.

Gould, H. A. 1960. Castes, outcastes, and the sociology of stratification. *International Journal of Comparative Sociology* 1: 220–38.

Gould, R. A. 1967. Notes on hunting, butchering, and sharing of game among the Ngatatjara and their neighbors in the West Australian Desert. *Kroeber Anthropological Society Papers* 36: 41–66.

Gould, R. A. 1980. *Living Archaeology*. Cambridge, Cambridge University Press.

Grandin, T. 2010. Auditing animal welfare at slaughter plants. *Meat Science* 86(1): 56–65.

Grant, A. 1971. The animal bones. In Cunliffe, B. (ed.), *Excavations at Fishbourne 1961–69*. London, Antiquities Research Report 26–7: 377–88.

Grant, A. 1975. The animal bones. In Cunliffe, B. (ed.), *Excavations at Porchester Castle*, vol. 1: *Roman. Reports of the Research Committee*. London, Society of Antiquaries of London: 378–408.

Grant, A. 1978. The animal bones. In Cunliffe, B. (ed.), *Excavations at Portchester Castle*, vol. III: *Medieval, the Outer Bailey and Its Defences*. London, Society of Antiquaries: 213–33.

Grant, A. 1987. Some observations of butchery in England from the Iron Age to medieval period. *Anthropozoologica*, Special (1): 53–9.

Grant, A. 1988. Animal resources. In Astill, G., & Grant, A. (eds.), *The Countryside of Medieval England*. Oxford, Basil Blackwell: 149–87.

Grant, A. 1989. Animals in Roman Britain. In Todd, M. (ed.), *Research on Roman Britain 1960–89*. London, Britannia Monograph Series, vol. 11: 135–47.

Grant, A. 2002. Food, status and social heirarchy. In Miracle, P., & Milner, N. (eds.), *Consuming Passions and Patterns of Consumption*. Cambridge, McDonald Institute for Archaeological Research: 17–25.

Grant, A. 2004. Domestic animals and their uses. In Todd, M. (ed.), *A Companion to Roman Britain*. London, Blackwell: 371–92.

Gravina. B, Rabbett, R., & Seetah, K. 2012. Combining stones and bones, defining form and function, inferring lives and roles. In Seetah, K., & Gravina, B. (eds.), *Bones for Tools – Tools for Bones*. Cambridge, McDonald Institute for Archaeological Research: 1–10.

Greenfield, H. 1999. The origins of metallurgy: distinguishing stone from metal cut-marks on bones from archaeological sites. *Journal of Archaeological Science* 26: 797–808.

Greenfield, H. 2000. The origins of metallurgy in the Central Balkans base on the analysis of cut marks on animal bones. *Environmental Archaeology* 5: 93–106.

Greenfield, H. 2006. Slicing cut marks on animal bones: diagnostics for identifying stone tool type and raw material. *Journal of Field Archaeology* 31(2): 147–63.

Greenfield, H. 2013a. 'The Fall of the House of Flint': a zooarchaeological perspective on the decline of chipped stone tools for butchering animals in the Bronze and Iron Ages of the southern Levant. *Lithic Technology* 38: 161–78.

Greenfield, H. 2013b. Monitoring the origins of metallurgy: an application of cut mark analysis on animals' bones from the Central Balkans. *Environmental Archaeology* 5(1): 93–106.

Greenfield, H., & Bouchnick, R. 2011. Kashrut and shechita: the relationship between dietary practices and ritual slaughtering of animals on Jewish identity. In Amundsen-Meyer, L., Engel, N., & Pickering, S. (eds.), *Identity Crisis: Archaeological Perspectives on Social Identity*. Proceedings of the 42nd Annual Chacmool Conference, University of Calgary, Calgary, Alberta, Canada: 106–21.

Greenfield, H., & Brown, A. 2016. 'Making the cut': changes in butchering technology and efficiency patterns from the Chalcolithic to modern Arab occupations at Tell Halif, Israel. In Marom, N., Yeshuran, R., Weissbrod, L., & Bar-Oz, G. (eds.), *Bones and Identity: Zooarchaeological Approaches to Reconstructing Social and Cultural Landscapes in Southwest Asia*. Oxford, Oxbow: 273–91.

Greenlee, J. J., & Greenlee, M. H. W. 2015. The transmissible spongiform encephalopathies of livestock. *Institute of Laboratory Animal Research Journal* 56(1): 7–25.

Greep, S. J. 1987. Use of bone, antler and ivory in the Roman and medieval periods. In

Watkinson, D., & Starling, K. (eds.), *Archaeo-logical Bone, Antler and Ivory*. London, UK Institute of Conservation of Historic and Artistic Works: 3–5.

Groemer, G. 2001. The creation of the Edo outcaste order. *Journal of Japanese Studies* 27(2): 263–93.

Guilday, J. E., Parmalee, P. W., & Tanner, D. P. 1962. Aboriginal butchering techniques at the Eschelman site (36LA12), Lancaster County, Pennsylvania. *Pennsylvania Archaeologist* 32: 59–83.

Guiry, E. J., Hillier, M., & Richards, M. P. 2015. Mesolithic dietary heterogeneity on the European Atlantic coastline: stable isotope insights into hunter-gatherer diet and subsistence in the Sado Valley, Portugal. *Current Anthropology* 56(3): 460–70.

Hagelberg, E., Hofreiter, M., & Keyser, C. 2015. Ancient DNA: the first three decades. *Philosophical Transactions of the Royal Society of London B: Biological Sciences* 370: 20130371.

Hall, A. 1989. Archaeological introduction. In O'Connor, T. P. (ed.), *Bones from Anglo-Scandinavian Levels at 16–22 Coppergate*: Vol. AY 15/3. York, Council for British Archaeology: 141–2.

Hall, R. A. 1999. 16–22 Coppergate. In Bond, J. M., & O'Connor, T. P. (eds.), *Bones from Medieval Deposits at 16–22 Coppergate and Other Sites in York*. York, Council for British Archaeology: 306–11.

Hall, R. A., & Hunter-Man, K. 2003. *Medieval Urbanism in Coppergate: Refining a Townscape*. York, CBA.

Hallpike, C. R. 1968. The status of craftsmen among the Konso of south-west Ethiopia. *Africa* 38(3): 258–69.

Hambleton, E. 2013. The life of things long dead: a biography of Iron Age animal skulls from Battlesbury Bowl, Wiltshire. *Cambridge Archaeological Journal* 23(3): 477–94.

Hambleton, E., & Maltby, J. M. 2004. *Animal Bones from Medieval Contexts at Laugharne Castle, Dyfed, Wales*. Bournemouth: Unpublished Consultancy Report for CADW.

Hambleton, E., & Maltby, J. M. 2009. *Animal Bones from Caerwent Forum-Basilica, Wales*. Bournemouth: Unpublished Consultancy Report for National Museum of Wales.

Hammon, A. 2011. Understanding the Romano-British–early medieval transition: a zooarchaeological perspective from Wroxeter (*Viroconium Cornoviorum*). *Britannia* 42: 275–305.

Hanks, B. 2010. Archaeology of the Eurasian steppes and Mongolia. *Annual Review of Anthropology* 39: 469–86.

Hansson, A. 1996. *Chinese Outcasts: Discrimination and Emancipation in Late Imperial China*. London, Brill.

Harding, A. F. 2000. *European Societies in the Bronze Age*. Cambridge, Cambridge University Press.

Harris, M. 1985. *Good to Eat: Riddles of Food and Culture*. New York, NY, Simon and Schuster.

Hassig, D. 1995. *Medieval Bestiaries*. Cambridge, Cambridge University Press.

Haynes, G. 1991. *Mammoths, Mastodons, and Elephants: Biology, Behaviour and the Fossil Record*. Cambridge, Cambridge University Press.

Haynes, G., & Krasinski, K. E. 2010. Taphonomic field-work in Southern Africa and its application in studies of the earliest peopling of North America. *Journal of Taphonomy* 8(2–3): 181–202.

Hesse, R. 2011. Reconsidering animal husbandry and diet in the northwest provinces. *Journal of Roman Archaeology* 24: 215–48.

Higgs, E. S., 1972. *Papers in Economic Prehistory*. Cambridge, Cambridge University Press.

Hill, A. 1979. Butchery and natural disarticulation: an investigatory technique. *American Antiquity* 44: 739–44.

Hilyard, E. J. W. 1957. Cataractonium, fort and town. *Yorkshire Archaeology Journal* 39: 224–65.

Himsworth, J. B. 1953. *The Story of Cutlery: From Flint to Stainless Steel*. London, Earnest Benn.

Hodder, I. 1982. *Symbols in Action: Ethnoarchaeological Studies of Material Culture*. Cambridge, Cambridge University Press.

Hodder, I. 1986. Digging for symbols in science and history: a reply. In *Proceedings of the Prehistoric Society* (vol. 52). Cambridge, Cambridge University Press: 352–6.

Hodder, I. 1990. *The Domestication of Europe: Structure and Contingency in Neolithic Societies.* Oxford, Blackwell.

Hodder, I. 1992. *Theory and Practice in Archaeology.* London, Routledge.

Hodder, I. 1997. 'Always momentary, fluid and flexible': towards a reflexive excavation methodology. *Antiquity* 71: 691–700.

Hodder, I. 2012. *Entangled: An Archaeology of the Relationships between Humans and Things.* Chichester, John Wiley & Sons.

Hodgson, G. W. I. 2002. Animal bones from the 1958–9 Bypass excavation (site 433). In Wilson, P. R. (ed.), *Cataractonium: Roman Catterick and Its Hinterland. Excavations and Research, 1958–1997.* York, CBA Research Reports: 415.

Holbrook, N. 1998. *Cirencester: The Roman Town Defences, Public Buildings and Shops.* Cirencester, Cotswold Archaeological Trust.

Hopkins, E. W. 1906. The Buddhistic rule against eating meat. *Journal of the American Oriental Society* 27: 455–64.

Horwitz, L. K., & Smith, P. 2000. The contribution of animal domestication to the spread of zoonoses: a case study from the southern Levant. *Anthropozoologica* 31: 77–84.

Horowitz, M. M. 1974. Barbers and bearers: ecology and ethnicity in an Islamic society. *Africa: Journal of the International African Institute* 44(4): 371–82.

Horowitz, R. 2006. *Putting Meat on the American Table: Taste, Technology.* Baltimore, MD, Johns Hopkins University Press.

Howell, S. 1996. Nature in culture or culture in nature? Chewong ideas of 'humans' and other species. In Descola, P., & Palsson, G. (eds.), *Nature and Society: Anthropological Perspectives.* London, Routledge: 127–45.

Hurst, H. 2005. Roman Cirencester and Gloucester compared. *Oxford Journal of Archaeology* 24(3): 293–305.

Ingold, T. 2000. *The Perception of the Environment: Essays on Livelihood, Dwelling and Skill.* London, Routledge.

Ireland, S. 2008. *Roman Britain: A Sourcebook.* London, Routledge.

Isaac, G. L. 1981. Archaeological tests of alternative models of early hominid behaviour: excavation and experiments. *Philosophical Transactions of the Royal Society of London B: Biological Sciences* 292(1057): 177–88.

Jackson, H. E. 2014. Animals as symbols, animals as resources. In Arbuckle, B. S., & McCarty, S. A. (eds.), *Animals and Inequality in the Ancient World.* Boulder, CO, University Press of Colorado: 107–25.

Jackson, H. E., & Scott, S. L. 2003. Patterns of elite faunal utilization at Moundville, Alabama. *American Antiquity* 68(3), 552–72.

James, S., & Millett, M. (eds.). 2001. *Britons and Romans: Advancing an Archaeological Agenda.* York, CBA Research Report, vol. 125.

Jayathilakan, K., Sultana, K., Radhakrishna, K., & Bawa, A. S. 2012. Utilization of byproducts and waste materials from meat, poultry and fish processing industries: a review. *Journal of Food Science and Technology* 49(3): 278–93.

Jodhka, S. S. 2012. *Caste.* Oxford, Oxford University Press.

Johnson, L. L. 1978. A history of flint-knapping experimentation, 1838–1976. *Current Anthropology* 19(2): 337–72.

Jones, A. 2001. *Archaeological Theory and Scientific Practice.* Cambridge, Cambridge University Press.

Jones, B. 2015. The butcher, vintner, and cidermaker: crafting good food in contemporary America. *Cuizine: The Journal of Canadian Food Cultures* 6(1), doi: 10.7202/1032255.

Jones, J. R. F. 1987. A false start? The Roman urbanization of Western Europe. *World Archaeology* 19: 47–57.

Jones, P. E. 1976. *The Butchers of London.* London, Secker & Warburg.

Jones, P. R. 1980. Experimental butchery with modern stone tools and its relevance for palaeolithic archaeology. *World Archaeology* 12(2): 153–65.

Jones, R. T. Langley, P., & Wall, S. 1985. The animal bones from the 1977 excavations. In Hinchcliffe, J., & Green, C. S. (eds.), *Excavations at Brancaster 1974 and 1977.* Norfolk, East Anglian Archaeological Report: 23.

Jørgensen, L. B. 2012. Writing craftsmanship? In Wendrich, W. (ed.), *Archaeology and Apprenticeship: Body Knowledge, Identity, and*

Communities of Practice. Tucson, AZ, University of Arizona Press: 240–54.

Judd, R. 2007. *Contested Rituals: Circumcision, Kosher Butchering, and Jewish Political Life in Germany, 1843–1933.* Ithaca, NY, Cornell University Press.

Kapp, L., Kapp, H., & Yoshihara, Y. 1987. *The Craft of the Japanese Sword.* Tokyo, Kodansha International.

Keller, C. M., & Keller, J. D. 1996. *Cognition and Tool Use: The Blacksmith at Work.* Cambridge, Cambridge University Press.

Kent, S. 1993. Sharing in an egalitarian Kalahari community. *Man* 28(3): 479–514.

Khare, R. S. 1966. A case of anomalous values in Indian civilisation: meat-eating among the Kanya-Kubja Brahmins of Kayayan Gotra. *Journal of Asian Studies* 25(2): 229–40.

Kim, J. S. 2013. *The Korean Paekjong under Japanese Rule: The Quest for Equality and Human Rights.* London, Routledge.

King, A. 1991. Food production and consumption: meat. In Jones, J. R. F. (ed.), *Roman Britain: Recent Trends.* Sheffield, J. R. Collis.

King, A. 1999. Diet in the Roman world: a regional inter-site comparison of the mammal bones. *Journal of Roman Archaeology* 12: 168–202.

Klein, R. G., & Cruz-Uribe, K. 1984. *The Analysis of Animal Bones from Archeological Sites.* Chicago, IL, University of Chicago Press.

Kleindienst, M. R., & Keller, C. M. 1976. Towards a functional analysis of handaxes and cleavers: the evidence from Eastern Africa. *Man* 11(2): 176–87.

Knappett, C. 2008. The neglected networks of material agency: artifacts, pictures, and texts. In Knappett, C., & Malafouris, L. (eds.), *Material Agency: Towards a Non-Anthropocentric Approach.* New York, NY, Springer: 139–56.

Kohn, M., & Mithen, S. 1999. Handaxes: products of sexual selection? *Antiquity* 73(281): 518–26.

Kristiansen, K., & T. Larsson, 2005. *The Rise of Bronze Age Society: Travels, Transmissions and Transformations.* Cambridge, Cambridge University Press.

Lahiri, N. 1995. Indian metal and metal-related artefacts as cultural signifiers: an ethnographic perspective. *World Archaeology* 27(1): 116–32.

Lapham, H. A., Balkansky, A. K., & Amadio, A. M. 2013. Animal use in the Mixteca Alta, Oaxaca, Mexico. In Emery, K. F., & Götz, C. M. (eds.), *The Archaeology of Mesoamerican Animals.* Atlanta, GA, Lockwood Press: 129–51.

Lartet, M. E. 1860. On the coexistence of man with certain extinct quadrupeds, proved by fossil bones, from various Pleistocene deposits, bearing incisions made by sharp instruments. *Quarterly Journal of the Geological Society* 16(1–2): 471–9.

Latour, B. 1987. *Science in Action: How to Follow Scientists and Engineers through Society.* Cambridge, MA, Harvard University Press.

Latour, B. 1996. On actor–network theory: a few clarifications. *Soziale Welt*: 369–81.

Latour, B. 1999. On recalling ANT. *The Sociological Review* 47(S1): 15–25.

Latour, B. 2005. *Reassembling the Social: An Introduction to Actor–Network Theory.* Oxford, Oxford University Press.

Lauwerier, R. C. 1988. *Animals in Roman Times in the Dutch Eastern River Area.* Amersfoort: Nederlandse Oudheden 12/Project Oostelijk Rivierengebied.

Law, J. 1984. On the methods of long-distance control: vessels, navigation and the Portuguese route to India. *The Sociological Review* 32(S1): 234–63.

Lawrence, S., & Davis. P. 2011. *An Archaeology of Australia since 1788.* New York, NY, Springer.

Lee, P. Y. 2008. *Meat, Modernity, and the Rise of the Slaughterhouse.* Durham, NH, University Press of New England.

Lehmann, G. 2003. The late medieval menu in England: a reappraisal. *Food & History* 1(1): 49–83.

LeMoine, G. 2002. Monitoring developments: replicas and reproducibility. In Mathieu, J. R. (ed.), *Experimental Archaeology: Replicating Past Objects, Behaviours, and Processes.* Oxford, Archaeopress, BAR S1035: 13–25.

Levine, M. 1986. Animal remains. In Hurst, H. R. (ed.), *Gloucester, the Roman and Later Defences.* Chichester, Gloucester Archaeological Repor: 81–4.

Lévi-Strauss, C. 1966. The culinary triangle. *New Society* 166: 937–40.

Lévi-Strauss, C. 1970. *The Raw and the Cooked*. London, Cape.

Lévi-Strauss, C. 1978. *The Origin of Table Manners*. London, Cape.

Levitan, B. 1984. Faunal remains from Priory Barn and Benham's Garage. In Leach, P. (ed.), *The Archaeology of Taunton: Excavations and Fieldwork to 1980*. Gloucester, Western Archaeological Trust Excavations Monograph, 8: 167–90.

Levy, T. E. 2007. *Journey to the Copper Age: Archaeology in the Holy Land*. San Diego, CA, San Diego Museum of Man.

Lewin, I., Munk, M., & Berman, J. 1946. *Religious Freedom: The Right to Practice Shehitah*. New York, NY, Research Institute for Post-War Problems of Religious Jewry.

Lewis, J. E. 2008. Identifying sword marks on bone: criteria for distinguishing between cut marks made by different classes of bladed weapons. *Journal of Archaeological Science* 35(7): 2001–8.

Lignereux, Y., & Peters, J. 1996. Techniques de boucherie et rejets osseux en Gaule Romaine. *Anthropozoologica* 24: 45–99.

Linduff, K. (ed.) 2004. *Metallurgy in Ancient Eastern Eurasia from the Urals to the Yellow River*. New York, NY, Edwin Mellon Press.

Lipschutz, Y. 1988. *Kashruth: A Comprehensive Background and Reference Guide to the Principles of Kashruth*. Brooklyn, NY, Mesorah Publications.

Loverdo, Jean de, 1906. *Les Abattoirs Publics: Construction etc Agencement des Abattoirs*, vol. 1. Paris.

Luncz, L. V., Wittig, R. M., & Boesch, C. 2015. Primate archaeology reveals cultural transmission in wild chimpanzees (*Pan troglodytes verus*). *Philosophical Transactions of the Royal Society of London B: Biological Sciences* 370 (1682), 20140348.

Lupo, K. D. 1994. Butchering marks and carcass acquisition strategies: distinguishing hunting from scavenging in archaeological context. *Journal of Archaeological Science* 21: 827–37.

Lupo, K. D. 2002. Cut and tooth mark distributions on large animal bones: ethnoarchaeological data from the Hadza and their implications for current ideas about early human carnivory. *Journal of Archaeological Science* 29: 85–109.

Lupo, K. D. 2006. What explains the carcass field processing and transport decisions of contemporary hunter-gatherers? Measures of economic anatomy and zooarchaeological skeletal part representation. *Journal of Archaeological Method and Theory* 13(1): 19–66.

Luzar, J. B., Silvius, K. M., & Fragoso, J. M. 2012. Church affiliation and meat taboos in indigenous communities of Guyanese Amazonia. *Human Ecology* 40(6): 833–45.

Lyman, R. L. 1977. Analysis of historic faunal remains. *Historical Archaeology* 11(1): 67–73.

Lyman, R. L. 1985. Bone frequencies: differential transport, in situ destruction, and the MGUI. *Journal of Archaeological Science* 12: 221–36.

Lyman, R. L. 1987. Archaeofaunas and butchery studies: a taphonomic perspective. In *Advances in Archaeological Methods and Theory*, vol. 10. London, Academic Press: 249–337.

Lyman, R. L. 1994. *Vertebrate Taphonomy*. Cambridge, Cambridge University Press.

Lyman, R. L. 2005. Analyzing cut marks: lessons from artiodactyl remains in the northwestern United States. *Journal of Archaeological Science* 32(12), 1722–32.

Lyman, R. L. 2010. What taphonomy is, what it isn't, and why taphonomists should care about the difference. *Journal of Taphonomy* 8(1): 1–16.

Lytton, T. D. 2013. *Kosher*. Cambridge, MA, Harvard University Press.

Machin, A. J., Hosfield, R., & Mithen, S. 2005. Testing the functional utility of axe symmetry: fallow deer butchery with replica hand-axes. *Lithics* 26: 23–38.

MacGregor, A. 2012. *Animal Encounters: Human and Animal Interaction in Britain from the Norman Conquest to World War I*. London, Reaktion Books.

MacGregor, A., Mainman, A. J., & Rogers, N. S. H. 1999. *Industry and Everyday Life: Bone, Antler, Ivory and Horn from Anglo-Scandinavian and Medieval York*. York, CBA.

MacKinnon, M. 2004. Production and consumption of animals in Roman Italy. *Journal of Roman Archaeology*, Supplement Series No. 54.

MacLachlan, I. 2008. Humanitarian reform, slaughter technology, and butcher resistance in nineteenth-century Britain. In Lee, P. Y. (ed.), *Meat, Modernity, and the Rise of the Slaughterhouse*. Durham, NH, University Press of New England: 107–27.

Mac Mahon, A. 2005. The shops and workshops of Roman Britain. In Mac Mahon, A., & Price, J. (eds.), *Roman Working Lives and Urban Living*. Oxford, Oxbow Books: 48–70.

Maddin, R., Muhly, J. D., & Stech, T. 1999. Early metalworking at Çayönü. In Hauptmann, A., Pernicka, E., Rehren, T., & Yalçin, Ü. (eds.), *The Beginnings of Metallurgy*. Bochum, Deutsches Bergbau-Museum: 36–44.

Maltby, J. M. 1979. *Faunal Studies on Urban Sites: The Animal Bones from Exeter*. Exeter, Exeter Archaeological Reports.

Maltby, J. M. 1981. Iron Age, Romano-British and Anglo-Saxon animal husbandry: a review of the faunal evidence. In Dimbleby, G. W., & Jones, M. (eds.), *The Environment of Man, the Iron Age to Anglo-Saxon Period*. Oxford, BAR British Series: 155–204.

Maltby, J. M. 1984. Animal bones and the Romano-British economy. In Grigson, C., & Clutton-Brock, J. (eds.), *Animals and Archaeology 4: Husbandry in Europe*, Oxford, BAR International Series: 125–37.

Maltby, J. M. 1985a. Assessing variations in Iron Age and Roman butchery practices: need for quantification. In Fieller, N. R., Gilbertson, D. D., & Ralph, N. G. (eds.), *Palaeobiological Investigations*. Oxford, BAR International Series: 19–30.

Maltby, J. M. 1985b. Patterns in faunal variability. In Barker, G. (ed.), *Beyond Domestication in Prehistoric Europe: Investigations in Subsistence Archaeology and Social Complexity*. New York, NY, Academic Press: 33–74.

Maltby, J. M. 1989. Urban and rural variation in the butchery of cattle in Romano-British Hampshire. In Serjeantson, D., & Waldron, T. (eds.), *Diets and Crafts in Towns*. Oxford, BAR British Series: 75–107.

Maltby, J. M. 1990. Animal bones from the excavations at 33 Sheep Street, Cirencester. In King, A. (ed.), *33 Sheep Street, Cirencester: Evaluation, Excavation, and Watching Brief*. CAT Typescript Report 9016: 34–52.

Maltby, J. M. 1993a. Animal bones. In Woodward, P., Davies, S. M., & Graham, A. (eds.), *Excavations at the Old Methodist Chapel and Greyhound Yard, Dorchester, 1981–84*. Dorchester, Dorset Natural History and Archaeology Society Monograph Series: 315–40.

Maltby, J. M. 1993b. The animal bones from a Roman-British well at Oakridge II, Basingstoke. *Proceedings of the Hampshire Field Club and Archaeological Society* 49: 47–76.

Maltby, J. M. 1994. The meat supply in Roman Dorchester and Winchester. In Hall, A., & Kenward, H. (eds.), *Urban-Rural Connexions: Perspectives from Environmental Archaeology*. Oxford, Oxbow Books: 85–102.

Maltby, J. M. 1996a. Animal bones. In Holbrook, N., & Thomas, A. (eds.), *Oxoniensia* 61: 158–60.

Maltby, J. M. 1996b. The exploitation of animals in the Iron Age: the archaeozoological evidence. In Champion, T., & Collis, J. (eds.), *The Iron Age in Britain and Ireland; Recent Trends*. Sheffield, J. R. Collis: 17–27.

Maltby, J. M. 1998a. Animal bones. *Transactions of the Bristol and Gloucestershire Archaeological Society* 116: 117–39.

Maltby, J. M. 1998b. Animal bones. In Barnes, I., Butterworth, C. A., Hawkes, J., & Smith, L. (eds.), *Excavations at Thames Valley Park, Reading, 1986–88. Prehistoric and Romano-British Occupation of the Floodplain and Terrace of the River Thames*. Salisbury, Trust for Wessex Archaeology: 69–72.

Maltby, J. M. 1998c. Animal bones from Romano-British deposits in Cirencester. In Holbrook, N. (ed.), *Cirencester: The Roman Town Defences, Public Buildings and Shops*. Cirencester, Cotswold Archaeological Trust: 352–70.

Maltby, J. M. 2007. Chop and change: specialist cattle carcass processing in Roman Britain. In Croxford, B., Ray, N., Roth, R., & White, N. (eds.), *TRAC 2006: Proceedings of the*

Sixteenth Annual Theoretical Roman Archaeology Conference, Cambridge 2006. Oxford, Oxbow: 59–76.

Maltby, J. M. 2010. *Feeding a Roman Town: Environmental Evidence from Excavations in Winchester, 1972–1985*. Winchester, Winchester Museums Service.

Maltby, J. M. 2013. The exploitation of animals in towns in the Medieval Baltic trading network: a case study from Novgorod. *Archaeology of the Baltic Region*. Saint Petersburg, Nestor-Historia: 229–44.

Maltby, J. M. 2014. Recording butchery and bone working. In Baker, P., & Worley, F. (eds.), *Animal Bones and Archaeology: Guidelines for Best Practice*. Portsmouth, English Heritage, 36–7.

Maltby, J. M. 2015. Commercial archaeology, zooarchaeology and the study of Romano-British towns. In Fulford, M., & Holbrook. N. (eds.), *The Towns of Roman Britain: The Contribution of Commercial Archaeology since 1990*. London, Britannia Monograph 27: 175–93.

Maltby, J. M. 2017. The exploitation of animals and their contribution to urban food supply in Roman southern England. In Bird, D. (ed.), *Agriculture and Industry in South-Eastern Roman Britain*. Oxford, Oxbow Books: 180–210.

Maltby, J. M. forthcoming. *Animal bones from Wortley*.

Mancini, R. A. 2009. Meat color. In Kerry, J., & Ledward, D. (eds.), *Improving the Sensory and Nutritional Quality of Fresh Meat*. Boca Raton, FL, CRC Press: 89–110.

Mancini, R. A., & Hunt, M. 2005. Current research in meat color. *Meat Science* 71(1): 100–21.

Manning, W. H. 1966. *A Hoard of Romano-British Ironworks from Brampton, Cumberland*. Transactions of the Cumberland & Westmoreland Antiquarian Society.

Manning, W. H. 1976. *Catalogue of Romano-British Ironworks in the Museum of Antiquities Newcastle upon Tyne*. Newcastle, University of Newcastle upon Tyne.

Manning, W. H. 1979. The native and Roman contribution to the development of metal industries in Britain. In Burnham, B. C., & Johnson, H. B. (eds.), *Invasion and Response*. BAR British Series, 73: 111–21.

Manning, W. H. 1985. *Catalogue of the Romano-British Iron Tools, Fittings and Weapons in the British Museum*. London, British Museum Publications.

Marchand, T. H. 2001. *Minaret Building and Apprenticeship in Yemen*. Surrey, Curzon.

Marchand, T. H. 2008. Muscles, morals and mind: craft apprenticeship and the formation of person. *British Journal of Educational Studies* 56(3): 245–71.

Marchand, T. H. 2010. Embodied cognition and communication: studies with British fine woodworkers. *Journal of the Royal Anthropological Institute* 16(s1): S100–S120.

Marciniak, A. 2005. *Placing Animals in the Neolithic: Social Zooarchaeology of Prehistoric Farming Communities*. London, University College London Press.

Marciniak, A., & Greenfield, H. 2013. A zooarchaeological perspective on the origins of metallurgy in the North European Plain: butchering marks on bones from Central Poland. In Bergerbrant, S., & Sabatini, S. (eds.), *Counterpoint: Essays in Archaeology and Heritage Studies in Honour of Professor Kristian Kristiansen*. Oxford, Archaeopress: 457–69.

Marcus, E. 2005. *Meat Market: Animals, Ethics and Money*. Boston, MA, Brio Press.

Martinón-Torres, M. 2002. Chaîne opératoire: the concept and its applications within the study of technology. *Gallaecia* 21: 29–44.

Masri, A. B. A. 1989. *Animals in Islam*. Petersfield, Athene Trust.

Mathieu, J. R. (ed.) 2002. *Experimental Archaeology: Replicating Past Objects, Behaviours, and Processes*. Oxford, Archaeopress, BAR S1035.

Matthews, S. 2005. The materiality of gesture: Intimacy, emotion and technique in the archaeological study of bodily communication. Position paper presented to the roundtable session 'The Archaeology of Gesture: Reconstructing Prehistoric Technical and Symbolic Behaviour' at the 11th Annual Meeting of the European Association of Archaeologists, 5–11 September, Cork, Ireland.

Mauss, M. 1973. Techniques of the body. *Economy and Society*, 2(1): 70–88.

Maxon, R. M. 1992a. The colonial financial system. In Ochieng, W. R., & Maxon, R. M. (eds.), *An Economic History of Kenya*. Nairobi, East African Publishers: 249–54.

Maxon, R. M. 1992b. Small-scale and large-scale agriculture since independence. In Ochieng, W. R., & Maxon, R. M. (eds.), *An Economic History of Kenya*. Nairobi, East African Publishers: 273–96.

McClain, J. 2002. *Japan: A Modern History*. New York, NY, W. W. Norton.

McCormack, N. Y. 2013. *Japan's Outcaste Abolition: The Struggle for National Inclusion and the Making of the Modern State*. London, Routledge.

McCormick, F. 1992. Early faunal evidence for dairying. *Oxford Journal of Archaeology* 11(2): 201–9.

McDonnell, G. 1989. Iron and its alloys in the fifth to eleventh centuries AD in England. *World Archaeology* 20(3): 373–82.

Mead, W. E. 1931. *The Medieval Feast*. London, George Allen and Unwin.

Meadow, R. H. 1978. "BONECODE": a system of numerical coding for faunal data from Middle Eastern sites. In Meadow, R. H., & Zeder, M. A. (eds.), *Approaches to Faunal Analysis in the Middle East*. Peabody Museum of Archaeology and Ethnology, Harvard University: 169–86.

Meddens, B. 1990a. Ancient Monuments Lab Rep 31/90, Yorkshire.

Meddens, B. 1990b. Ancient Monuments Lab Rep 98/90, Yorkshire.

Meddens, B. 2002. Animal bones from Bainesse. In Wilson, P. R. (ed.), *Cataractonium: Roman Catterick and Its Hinterland. Excavations and Research, 1958–1997*. York, CBA Research Reports, 129: 419.

Mendelsohn, O., & Vicziany, M. 1998. *The Untouchables: Subordination, Poverty and the State in Modern India* (vol. 4). Cambridge, Cambridge University Press.

Merritt, S. 2016. Cut mark cluster geometry and equifinality in replicated Early Stone Age butchery. *International Journal of Osteoarcheology* 26: 585–98.

Meskell, L. 2017. The archaeology of figurines and the human body. In Insoll, T. (ed.), *The Oxford Handbook of Prehistoric Figurines*. Oxford, Oxford University Press: 17–36.

Metcalfe, D., & Jones, K. T. 1988. A reconsideration of animal body-part utility indices. *American Antiquity* 53(3): 486–504.

Metcalfe, R. S. 2012. *Meat, Commerce and the City: The London Food Market, 1800–1855*. London, Pickering & Chatto.

Mettler, J. 1987. *Basic Butchering of Livestock and Game*. North Adams, MA, Storey Books.

Miles, D. 2016. *The Tale of the Axe: How the Neolithic Revolution Transformed Britain*. London, Thames & Hudson.

Miller, D. 1995. Consumption and commodities. *Annual Review of Anthropology* 24: 141–61.

Miller, H. M. L. 2012. Types of learning in apprenticeship. In Wendrich, W. (ed.), *Archaeology and Apprenticeship: Body Knowledge, Identity and Communities of Practice*. Tuscon, AZ, University of Arizona Press: 224–39.

Mines, D. P. 2009. *Caste in India*. Ann Arbor, MI, Association for Asian Studies.

Minnich, R. G. 1990. The gift of koline and the articulation of identity in Slovene peasant society. *Etnologia Slavica* 22: 151–61.

Miracle, P. 2002. Mesolithic meals from Mesolithic middens. In Miracle, P., & Milner, N. (eds.), *Consuming Passions and Patterns of Consumption*. Cambridge, McDonald Institute for Archaeological Research: 65–88.

Mooketsi, C. 2001. Butchery styles and the processing of cattle carcasses in Botswana. *PULA Botswana Journal of African Studies* 15(1): 108–24.

Moore, R., Stone, J., & Tattersall, H. 1983. *The Meat Buyers Guide for Caterers*. London, Northwood Books.

Moore, S. F. 1894. The butchers of Korea. *Korean Report* 5: 127–32.

Morales, J., Rodríguez, A., Alberto, V., Machado, C., & Criado, C. 2009. The impact of human activities on the natural environment of the Canary Islands (Spain) during the pre-Hispanic stage (3rd–2nd century BC to 15th century AD): an overview. *Environmental Archaeology* 14(1): 27–36.

Morgan, W. T. W. 1963. The 'white highlands' of Kenya. *The Geographical Journal* 129(2): 140–55.

Mortimer, R., Regan, R., & Lucy, S. 2005. *Saxon and Medieval Settlement at West Fen Road, Ely: The Ashwell Site.* Cambridge, East Anglia Archaeology Report 110.

Motoyama, M., Sasaki, K., & Watanabe, A. 2016. Wagyu and the factors contributing to its beef quality: a Japanese industry overview. *Meat Science* 120: 10–18.

Mowat, F. 1970. *People of the Deer.* Toronto, McClelland & Stewart.

Muga, G. O., Onyango-Ouma, W., Sang, R., & Affognon, H. 2015. Sociocultural and economic dimensions of Rift Valley fever. *American Journal of Tropical Medicine and Hygiene* 92 (4): 730–8.

Muhly, J. D. 2013. Metals and metallurgy. In Steadman, S. R., & McMahon, G. (eds.), *The Oxford Handbook of Ancient Anatolia 10,000–323 B.C.E.* Oxford, Oxford University Press: 858–77.

Murphy, P., Albarella, U., Germany, M., & Locker, A. 2000. Production, imports and status: biological remains from a late Roman farm at Great Holts Farm, Boreham, Essex, UK. *Environmental Archaeology* 5: 35–48.

Murray, E., McCormick, F., & Plunkett, G. 2004. The food economies of Atlantic island monasteries: the documentary and archaeo-environmental evidence. *Environmental Archaeology* 9(2): 179–88.

Nakamura, C., & Meskell, L. 2009. Articulated bodies: forms and figures at Çatalhoöyük. *Journal of Archaeological Methods and Theory* 16: 205–30.

Natrajan, B. 2012. *The Culturalization of Caste in India: Identity and Inequality in a Multicultural Age.* London, Routledge.

Navarrete, A., van Schaik, C. P., & Isler, K. 2011. Energetics and the evolution of human brain size. *Nature* 480(7375): 91–3.

Neary, I. 1987. The Paekjong and the Hyong-pyongsa: the Untouchables of Korea and their struggle for liberation. *Immigrants & Minorities* 6(2): 117–50.

Noddle, A. 1984. A comparison of the bones of cattle, sheep and pigs from ten Iron Age and Romano-British sites. In Grigson, C., & Clutton-Brock, J. (eds.), *Animals and Archaeology 4: Husbandry in Europe.* Oxford, BAR Publishing: 105–23.

Noddle, A. 1989. Flesh on the bones: some notes on animal husbandry in the past. *Archaeozoologia* 3(1–2): 25–50.

Noddle, A. 2000. Large vertebrate remains. In Price, E. (ed.), *Frocester: A Romano-British Settlement, Its Antecedents and Successors.* Gloucester, Gloucester and District Archaeology Research Group: 217–43.

Noe-Nygaard, N. 1977. Butchering and marrow fracturing as a taphonomic factor in archaeological deposits. *Paleobiology* 3: 218–37.

Novak, V. 1970. *Živinoreja.* In Blaznik, P., Grafenauer, G., & Vilfan, S. (eds.), *Gospodarska in Družbena Zgodovina Slovencev: Zgovina agrarnih Panog – I. Zvezek: Agrarno Gospodarstvo.* Ljubljana, Državna Založba Slovenije.

Nozaki, H. 2009. *Japanese Kitchen Knives: Essential Techniques and Recipes.* Tokyo, Kodansha International.

Ocejo, R. E. 2014. Show the animal: constructing and communicating new elite food tastes at upscale butcher shops. *Poetics* 47: 106–21.

Ochieng, W. R. 1992. The post-colonial state and Kenya's economic inheritance. In Ochieng, W. R., & Maxon, R. M. (eds.), *An Economic History of Kenya.* Nairobi, East African Publishers: 259–72.

O'Connell, J. F., & Marshall, B. 1989. Analysis of kangaroo body part transport among the Alyawara of Central Australia. *Journal of Archaeological Science* 16(4): 393–405.

O'Connor, T. P. 1984. *Selected Groups of Bones from Skeldergate and Walmgate.* York, CBA.

O'Connor, T. P. 1988. Bones from the General Accident Site – Tanner Row. *Archaeology of York,* 15: 61–136.

O'Connor, T. P. 1989a. What shall we have for dinner? Food remains from urban sites. In Serjeantson, D., & Waldron, T. (eds.), *Diets and Crafts in Towns,* Oxford, BAR British Series: 13–23.

O'Connor, T. P. 1989b. *Bones from Anglo-Scandinavian Levels at 16–22 Coppergate.* York, CBA.

O'Connor, T. P. 1991. *Bones from 46–54 Fishergate*. York, CBA.

O'Connor, T. P. 1993. Bone assemblages from monastic sites: many questions but few data. In Gilchrist, R., & Mytum, H. (eds.), *Advances in Monastic Archaeology*. Oxford, BAR British Series: 107–11.

O'Connor, T. P. 2005. *Biosphere to Lithosphere: New Studies in Vertebrate Taphonomy*. Oxford, Oxbow Books.

O'Connor, T. P. 2008. *The Archaeology of Animal Bones*. College Station, Texas A&M University Press.

O'Connor, T. P. 2014. Livestock and animal husbandry in early medieval England. *Quaternary International* 346: 109–18.

O'Day, S. J., Van Neer, W., & Ervynck, A. 2004. *Behaviour behind Bones*. Oxford, Oxbow Books.

Ogonda, R. T., & Ochieng, W. R. 1992. Land, natural and human resources. In Ochieng, W. R., & Maxon, R. M. (eds.), *An Economic History of Kenya*. Nairobi, East African Publishers: 1–16.

Oliphant, G. G. 1997. Meat and meat products. In Ranken, M. D., Kill, R. C., & Baker, C. G. J. (eds.), *Food Industries Manual*. London, Blackie Academic and Professional: 28–30.

Olsen, B. 2003. Material culture after text: remembering things. *Norwegian Archeological Review* 36(2): 87–104.

Olsen, S. 1989. On distinguishing natural from cultural damage on archaeological antler. *Journal of Archaeological Science* 16: 125–35.

Olsen, S., & Shipman, P. 1988. Surface modification on bone: trampling versus butchery. *Journal of Archaeological Science* 15: 535–53.

Orton, D. C. 2010. A new tool for zooarchaeological analysis: ArcGIS skeletal templates for some common mammalian species. *Internet Archaeology* (28), http://dx.doi.org/10.11141/ia.28.4.

Orton, D. C. 2012. Taphonomy and interpretation: an analytical framework for social zooarchaeology. *International Journal of Osteoarchaeology* 22(3): 320–37.

Otter, C. 2008. Civilizing slaughter: the development of the British public abattoir, 1850–1910. In Lee, P. Y. (ed.), *Meat, Modernity, and the Rise of the Slaughterhouse*. Durham, NH, University Press of New England: 89–107.

Outram, A. 2002. Bone fracture and within-bone nutrients: an experimentally based method for investigating levels of marrow extraction. In Miracle, P., & Milner, N. (eds.), *Consuming Passions and Patterns of Consumption*. Cambridge, McDonald Institute for Archaeological Research: 51–65.

Outram, A., & Rowley-Conwy, P. 1998. Meat and marrow utility indices for horse (*Equus*). *Journal of Archaeological Science* 25: 839–49.

Pacyga, D. A. 2015. *Slaughterhouse: Chicago's Union Stock Yard and the World It Made*. Chicago, IL, University of Chicago Press.

Padrón, F. M. 1993. *Canarias. Crónicas de su conquista 1500/1525*. Las Palmas de Gran Canari, Cabildo Insular de Gran Canaria.

Page, T. 1999. *Buddhism and Animals: A Buddhist Vision of Humanity's Rightful Relationship with the Animal Kingdom*. London, UKAVIS Publications.

Pálsson, G. 1994. Enskilment at sea. *Man* 29(4): 901–27.

Pankhurst, A. 1999. 'Caste' in Africa: the evidence from south-western Ethiopia reconsidered. *Africa* 69(4): 485–509.

Passin, H. 1955. Untouchability in the Far East. *Monumenta Nipponica* 11(3): 247–67.

Passin, H. 1957. The Paekchŏng of Korea: a brief social history. *Monumenta Nipponica* 12 (3–4): 195–240.

Payne, S. 1990. Ancient Monuments Lab Rep 5/90, Yorkshire.

Pearsall, J. 2016. *Concise Oxford English Dictionary*. Oxford, Oxford University Press.

Peck, R. W. 1986. Applying contemporary analogy to the understanding of animal processing behaviour on Roman villa sites. PhD dissertation, University of Southampton.

Pelegrin, J. 1990. Prehistoric lithic technology: some aspects of research. *Archaeological Review from Cambridge* 9(1): 116–25.

Perren, R. 2008. Filth and profit, disease and health: public and private impediments to slaughterhouse reform in Victorian Britain.

In Lee, P. Y. (ed.), *Meat, Modernity, and the Rise of the Slaughterhouse*. Durham, NH, University Press of New England: 127–50.

Pilcher, J. M. 2006. *The Sausage Rebellion: Public Health, Private Enterprise, and Meat in Mexico City, 1890–1917*. Albuquerque, NM, University of New Mexico Press.

Pimentel, D. 2004. Livestock production and energy use. In Cleveland, C. J. (ed.), *Encyclopedia of Energy*. San Diego, CA, Elsevier: 671–6.

Pineda, A., Saladié, P., Vergès, J., Huget, R., Cáceres, I., & Vallverdú, J. 2014. Trampling versus cut marks on chemically altered surfaces: an experimental approach and archaeological application at the Boella site (la Canonja, Tarragone, Spain). *Journal of Archaeological Science* 50: 84–93.

Pluskowski, A. 2007. Communicating through skin and bone: appropriating animal bodies in medieval Western European seigneurial culture. In Pluskowski, A. (ed.), *Breaking and Shaping Beastly Bodies: Animals as Material Culture in the Middle Ages*. Oxford, Oxbow: 35–51.

Pluskowski, A. 2010. The zooarchaeology of medieval 'Christendom': ideology, the treatment of animals and the making of medieval Europe. *World Archaeology* 42(2): 201–14.

Politis, G. C., & Saunders, N. J. 2002. Archaeological correlates of ideological activity: food taboos and spirit-animals in an Amazonian hunter-gatherer society. In Miracle, P., & Milner, N. (eds.), *Consuming Passions and Patterns of Consumption*. Cambridge, McDonald Institute for Archaeological Research: 113–31.

Pollan, M. 2008. *In Defense of Food: An Eater's Manifesto*. New York, NY, Penguin Books.

Popkin, P. 2005. Caprine butchery and bone modification templates: A step towards standardisation. *Internet Archaeology* 17, http://dx.doi.org/10.11141/ia.17.2.

Preucel, R. 2012. Archaeology and the limitations of actor network theory. Paper presented to the Department of Anthropology, Harvard University, 10 October.

Price, J. 1966. A history of the outcaste: untouchability in Japan. In De Vos, George A., & Wagatsuma, Hiroshi (eds.), *Japan's Invisible Race: Caste in Culture and Personality*. Berkeley, CA, University of California Press: 6–30.

Price, J. 2003. A history of the outcaste: untouchability in Japan. In Reilly, K., Kaufman, S., & Bodino, A. (eds.), *Racism: A Global Reader*. Armonk, NY, M. E. Sharpe: 38–42.

Purnell, G., Maddock, N. A., & Khodabandehloo, K. 1990. Robot deboning for beef forequarters. *Robotica* 8(4): 303–10.

Raab, L. M., & Goodyear, A. C. 1984. Middle-range theory in archaeology: A critical review of origins and applications. *American Antiquity* 49(2): 255–68.

Ranken, M. D. 2000. *Handbook of Meat Production Technology*. London, Blackwell Scientific.

Rawcliffe, C. 2013. *Urban Bodies: Communal Health in Late Medieval English Towns and Cities*. Woodbridge, Boydell & Brewer.

Renfrew, C. 1994. Towards a cognitive archaeology. In Renfrew, C., & Zubrov, E. (eds.), *The Ancient Mind: Elements of Cognitive Archaeology*. Cambridge, Cambridge University Press: 3–12.

Reitz, E., & Wing, E. 2000. *Zooarchaeology*. Cambridge, Cambridge University Press.

Ritvo, H. 1987. *The Animal Estate: The English and Other Creatures in the Victorian Age*. Cambridge, MA, Harvard University Press.

Rixson, D. 1988. Butchery evidence on animal bones. *Circaea* 6(1): 49–62.

Rixson, D. 2000. *The History of Meat Trading*. Nottingham, Nottingham University Press.

Robb, J. 2002. Time and biography. In Hamilakis, Y., Pluciennik, M., & Tarlow, S. (eds.), *Thinking through the Body: Archaeologies of Corporeality*. New York, NY, Springer: 153–71.

Romans, J. R., Costello, W. J., Jones, K. W., Carlson, C. W., & Ziegler, P. T. 1994. *The Meat We Eat*. Fort Lee, NJ, Interstate Printers and Publishers.

Ross, E. B. 1978. Food taboos, diet, and hunting strategy: the adaptation to animals in Amazon cultural ecology. *Current Anthropology* 19(1): 1–36.

Roux, V. 1989. *The Potter's Wheel Craft Specialization and Technical Competence*. New Delhi, Oxford University Press and IBH Publishing.

Rozin, P., Markwith, M., & Stoess, C. 1997. Moralization and becoming a vegetarian: the transformation of preferences into values and the recruitment of disgust. *Psychological Science* 8: 67–73.

Ruby, M. B., & Heine, S. J. 2011. Meat, morals, and masculinity. *Appetite* 56: 447–50.

Russell, M. D. 1987. Mortuary practice at the Krapina Neanderthal site. *American Journal of Physical Anthropology* 72: 381–97.

Russell, N. 2012. *Social Zooarchaeology: Humans and Animals in Prehistory*. Cambridge, Cambridge University Press.

Sabban, F. 1993. La viande en Chine: imaginaire et usages culinaires. *Anthropozoologica* 18: 79–90.

Sabine, E. L. 1933. Butchering in mediaeval London. *Speculum* 8(3): 335–53.

Sadek-Kooros, H. 1972. Primitive bone fracturing: a method of research. *American Antiquity* 37(3): 369–82.

Sahay, A., Tiwari, R., Roy, R., & Sharma, M. C. 2010. Study on meat associated health hazards among butchers and meat retailers. *Journal of Veterinary Public Health* 8(1): 33–6.

Salway, P. 2015. *Roman Britain: A Very Short Introduction*. Oxford, Oxford University Press.

Saul, F., & Saul, J. 1989. Osteobiography: a Maya example. In İşcan, M., & Kennedy, K. (eds.), *Reconstruction of Life from the Skeleton*. New York, NY, Wiley-Liss: 287–302.

Saunders, E. P. 1990. *Jewish Law from Jesus to Mishnam: Five Studies*. Philadelphia, PA, Trinity Press International.

Saunders, P. 2006. *Social Class and Stratification*. London, Routledge.

Savelle, J. M., & Friesen, T. M. 1996. An odontocete (Cetacea) meat utility index. *Journal of Archaeological Science* 23: 713–21.

Savelle, J. M. Friesen, T. M., & Lyman, R. L. 1996. Derivation and application of the otariid utility index. *Journal of Archaeological Science* 23: 705–12.

Schänzle, M. 1981. *Studies in the Breed History of the Rottweiler*. Hollywood, CA, Powderhorn Press.

Schiffer, M. B. 1975. Behavioral chain analysis: activities, organization, and the use of space. *Fieldiana Anthropology* 65: 103–19.

Schiffer, M. B. 1992. *Technological perspectives on behavioral change*. Tuscon, AZ, University of Arizona Press.

Schiffer, M. B. 2004. Studying technological change: a behavioral perspective. *World Archaeology* 36(4): 579–85.

Schlanger, N. 1994. Mindful technology: unleashing the chaîne opératoire for an archaeology of mind. In Renfrew, C., & Zubrov, E. (eds.), *The Ancient Mind: Elements of Cognitive Archaeology*. Cambridge, Cambridge University Press: 143–51.

Schmidt, C. W., Moore, C. R., & Leifheit, R. 2012. A preliminary assessment of using a white light confocal imaging profiler for cut mark analysis. In Bell, L. (ed.), *Forensic Microscopy for Skeletal Tissues: Methods and Protocols*. New York, Humana Press: 235–48.

Schmitz, P. 1945. *Sancti Benedicti Regula Monachorum*. Maredous: Editions de Maresdsous.

Seccombe, Rev. J. 1743. *Business and Diversion, Inoffensive to God and Necessary to the Comfort and Support of Human Society*. Boston, MA, S. Kneeland & T. Green.

Seetah, K. 2002. Techniques and implement use in urban Romano-British cattle butchery. MSc. thesis, School of Conservation Sciences. Bournemouth University.

Seetah, K. 2004. Meat in history – the butchery trade in the Romano-British Period. *International Journal of Food History* 2(2): 19–35.

Seetah, K. 2005. Butchery as a tool for understanding the changing views of animals. In Pluskowski, A. (ed.), *Just Skin and Bones? New Perspectives on Human–Animal Relations in the Historic Past*. Oxford, BAR International Series S1410: 1–8.

Seetah, K. 2006. Multidisciplinary approach to Romano-British cattle butchery. In Maltby, M. (ed.), *Integrating Zooarchaeology*. Oxford, Oxbow: 111–18.

Seetah, K. 2007. The Middle Ages on the block: animals, guilds and meat in medieval Britain. In Pluskowski, A. (ed.), *Breaking and Shaping Beastly Bodies: Animals as Material Culture in the Middle Ages*. Oxford, Oxbow: 18–31.

Seetah, K. 2008. Modern analogy, cultural theory and experimental archaeology: a merging point at the cutting edge of archaeology. *World Archaeology* 40(1): 135–50.

Seetah, K. 2010. Religion, legislation, and meat: the politics of food and its implications for the butchers of London. In Pluskowski, A. G., Kunst, G. K., Kucera, K., Bietak, M., & Hein, I. (eds.), *Bestial Mirrors: Using Animals to Construct Human Identities in Medieval Europe*. Vienna, Vias, vol. 3: 110–15.

Seetah, K. in prep a. Cracking the butchery code II: methodological approaches to cut mark analysis.

Seetah, K. in prep b. To each their own: The primacy of agency in decision making for butchery in Maasai culture and the relevance for understanding past practice.

Seetah, K. in prep c. Cracking the butchery code II: experimental replication of tools and techniques.

Seetah, K., Pluskowski, A., Makowiecki, D., & Daugnora, L. 2014. New technology or adaptation at the frontier? Butchery as a signifier of cultural transitions in the medieval Eastern Baltic. In Pluskowski, A. (ed.), *Life at the Frontier: The Ecological Signatures of Human Colonisation in the North*. Klaipėda, Archaeologica BALTICA, vol 20: 59–77.

Sellet, F. 1993. Chaîne opératoire: the concept and its applications. *Lithic Technology* 18(1–2): 106–12.

Serjeantson, D. 1989. Animal remains and the tanning trade. In Serjeantson, D., & Waldron, T. (eds.), *Diets and Crafts in Towns*. Oxford, BAR British Series: 129–43.

Shack, W. A. 1964. Notes on occupational castes among the Gurage of south-west Ethiopia. *Man* 64: 50–2.

Shah, G. (2006). *Untouchability in Rural India*. New Delhi, Sage.

Shanks, M. 2007. Symmetrical archaeology. *World Archaeology* 39(4): 589–96.

Shanks, M., & McGuire, R. H. 1996. The craft of archaeology. *American Antiquity* 61(1): 75–88.

Shipman, P. 1981. Application of scanning electron microscopy to taphonomic problems. *Annals of the New York Academy of Science* 276: 357–85.

Shipman, P. 1986. Scavenging or hunting in early hominids: theoretical framework and test. *American Anthropologist* 88: 27–43.

Shipman, P., & Rose, J. 1983. Early hominid hunting, butchering and carcass processing behaviour: approaches to the fossil record. *Journal of Anthropological Archaeology* 2: 57–98.

Siddiqi, A. 2001. Ayesha's world: a butcher's family in nineteenth-century Bombay. *Comparative Studies in Society and History* 43: 101–29.

Sidéra, I. 2013. Manufacturing bone tools: the example of Kovačevo. In Miladinović-Radmilović, N., & Vitezović, S. (eds.), *Bioarheologija na Balkanu. Bilans i perspektive. Radovi bioarheološke sekcije Srpskog arheološkog društva*. Beograd, Papers of the Bioarchaeological Section of the Serbian Archaeological Society: 173–9.

Sillar, B., & Tite, M. S. 2000. The challenge of 'technological choices' for materials science approaches in archaeology. *Archaeometry* 42(1): 2–20.

Sinclair, U. 1906. *The Jungle*. New York, NY, New American Library.

Sitz, B. M., Calkins, C. R., Feuz, D. M., Umberger, W. J., & Eskridge, K. M. (2006). Consumer sensory acceptance and value of wet-aged and dry-aged beef steaks. *Journal of Animal Science* 84(5): 1221–6.

Skibo, J. M., & Schiffer, M. 2008. *People and Things: A Behavioral Approach to Material Culture*. New York, NY, Springer Science & Business Media.

Smerdel, I. 2002. Slovenia: an enigma to Europeans (and others), Europe in a nutshell to Slovenes. In Lysaght, P. (ed.), *Food and Celebration from Fasting to Feasting*. Ljubljana, ZRC Publishing: 29–39.

Smith, A. L. 2000. Processing clay for pottery in northern Cameroon: social and technical requirements. *Archaeometry* 42(1): 21–42.

Spencer, H. 1898–1900. *The Principles of Sociology*, 3rd edn. New York, NY, D. Appleton.

Speth, J. D. 1983. *Bison Kills and Bone Counts: Decision Making by Ancient Hunters*. Chicago, IL, University of Chicago Press.

Speth, J. D. 1991. Taphonomy and early hominid behavior: problems in distinguishing cultural and non-cultural agent. In Stiner, M. C. (ed.), *Human Predators and Prey Mortality*. Boulder, CO, Westview Press: 31–40.

Spiess, A. E. 1979. *Reindeer and Caribou Hunters: An Archaeological Study*. New York, NY, Academic Press.

Stallibrass, S. 1997. Ancient Monuments Lab Rep 104/97, Yorkshire.

Stallibrass, S. 2002. An overview of the animal bones: what would we like to know, what do we know so far, and where do we go from here? In Wilson, P. R. (ed.), *Cataractonium: Roman Catterick and Its Hinterland. Excavations and Research, 1958–1997*. York, CBA Research Reports, 129: 392–415.

Stallibrass, S., & Thomas, R. (eds). 2008. *Feeding the Roman Army: The Archaeology of Production and Supply in North-West Europe*. Oxford: Oxbow.

Stanford, D. 1999. *The Hunting Apes*. Princeton, NJ, Princeton University Press.

Stanford, D., Bonnichsen, R., & Morlan, R. 1981. The Ginsberg experiment: modern and prehistoric evidence of bone flaking technology. *Science* 212: 434–40.

Staples, J. 2008. 'Go on, just try some!': meat and meaning-Making among South Indian Christians. *South Asia: Journal of South Asian Studies* 31(1): 36–55.

Sterner, J., & David, N. 1991. Gender and caste in the Mandara highlands: northeastern Nigeria and northern Cameroon. *Ethnology* 30(4): 355–69.

Stokes, P. 2000. A cut above the rest? Officers and men at South Shields Roman fort. In Rowley-Conwy, P. (ed.), *Animal Bones, Human Societies*. Oxford, Oxbow Books: 145–52.

Stout, D., & Chaminade, T. 2007. The evolutionary neuroscience of tool making. *Neuropsychologia* 45(5), 1091–100.

Sutton, P. C., Wieseman, M. E., Museum of Fine Arts, Boston, & Toledo Museum of Art. 1993. *The Age of Rubens*. Boston, Museum of Fine Arts in Association with Ghent.

Swatland, H. J. 2000. *Meat Cuts and Muscle Foods*. Nottingham, Nottingham University Press.

Swift, J. 1843. *The Works of Jonathan Swift*. London, H. G. Bohn.

Sykes, N. 2007. Taking sides: the social life of venison in medieval England. In Pluskowski, A. (ed.), *Breaking and Shaping Beastly Bodies: Animals as Material Culture in the Middle Ages*. Oxford, Oxbow: 149–60.

Sykes, N. 2014. *Beastly Questions: Animal Answers to Archaeological Issues*. London, Bloomsbury.

Testart, A. 1987. Game sharing systems and kinship systems among hunter-gatherers. *Man* 22 (2): 287–304.

Thawley, C. R. 1982. The animal remains. In Wacher, J., & Whirr, A. D. (eds.), *Early Roman Occupation at Cirencester*. Cirencester, Cirencester Evacuation Committee: 211–27.

Thomas, K. 1983. *Man and the Natural World: Changing Attitudes in England 1500–1800*. London, Allen Lane.

Thomas, P. H. 1963. Graeco-Roman medical and surgical instruments with special reference to Wales and the Border. *Journal of the Royal College of General Practitioners* 6(3): 495–502.

Thomas, R. 1999. Feasting at Worcester Cathedral in the 17th century: a zooarchaeological and historical investigation. *Archaeological Journal* 156: 342–58.

Thomas, R. 2005. Zooarchaeology, improvement and the British agricultural revolution. *International Journal of Historical Archaeology* 9 (2): 71–88.

Thomas, R. 2006. Chasing the ideal? Ritualism, pragmatism and the later medieval hunt in England. In Pluskowski, A. (ed.), *Breaking and Shaping Beastly Bodies: Animals as Material Culture in the Middle Ages*. Oxford, Oxbow: 125–48.

Tilley, C. 1994. Interpreting material culture. In Pearce, S. M. (ed.), *Interpreting Objects and Collections*. London, Routledge: 67–75.

Tobach, E. 1995. The uniqueness of human labor. In Martin, L. M., Nelson, K., & Tobach, E. (eds.), *Sociocultural Psychology: Theory and Practice of Doing and Knowing*. Cambridge, Cambridge University Press: 43–67.

Tobert, N. 1985. Craft specialisation: a seasonal camp in Kebkebiya. *World Archaeology* 17(2): 278–88.

Todd, D. M. 1977. Caste in Africa? *Africa* 47(4): 398–412.

Todd, D. 1978. The origins of outcastes in Ethiopia: reflections on an evolutionary theory. *Abbay* 9: 145–58.

Todd, L., & Rapson, D. J. 1988. Long bone fragmentation and interpretation of faunal assemblages: approaches to comparative analysis. *Journal of Archaeological Science* 15: 307–25.

Trigger, B. G. 1995. Expanding middle-range theory. *Antiquity* 69(264): 449–58.

Tringham, R. 1978. Experimentation, ethnoarchaeology, and the leapfrogs, in archaeological methodology. In Gould, R. A. (ed.), *Explorations in Ethnoarchaeology*. Albuquerque, University of New Mexico Press: 169–99.

Twiss, K. (ed). 2007. *The Archaeology of Food and Identity*. Carbondale, IL, Southern Illinois University. Occ. Paper 34.

Tylecote, R. F. 1986. *The Prehistory of Metallurgy in the British Isles*. London, Institute of Metals.

Tylecote, R. F. 1987. *The Early History of Metallurgy in Europe*. London, Longman.

Tylecote, R. F. 1992. *A History of Metallurgy*. London, Institute of Metals.

Valenzuela-Lamas, S., Valenzuela-Suau, L., Saula, O., Colet, A., Mercadal, O., Subiranas, C., & Nadal, J. 2014. Shechita and kashrut: identifying Jewish populations through zooarchaeology and taphonomy. Two examples from medieval Catalonia (north-eastern Spain). *Quaternary International* 330: 109–17.

Valerie, V. 2000. *The Forest of Taboos: Morality, Hunting and Identity among the Huaulu of Moluccas*. Madison, WI, University of Wisconsin Press.

Van-Mensch, P. J. A. 1974. A Roman soup kitchen at Zwammerdam? *Berichten van de Rijkdienst voor het Oudheidkundig Bodemonderzoek* 24: 159–65.

Vann, S., & Thomas, R. 2006. Humans, other animals and disease: a comparative approach towards the development of a standardised recording protocol for animal palaeopathology. *Internet Archaeology* 20, http://dx.doi.org/10.11141/ia.20.5.

Van Oyen, A. 2015. Actor–network theory's take on archaeological types: becoming, material agency and historical explanation. *Cambridge Archaeological Journal* 25(1): 63–78.

van Wijngaarden-Bakker, L. H. 1990. Replication of butchery marks on pig mandibles. In Robinson, D. E. (ed.), *Experiment and Reconstruction in Environmental Archaeology*. Oxford, Oxbow Books: 167–74.

Varela, F. J., Thompson, E. & E. Rosch. 1991. *The Embodied Mind: Cognitive Science and Human Experience*. Cambridge, MA, MIT Press.

Veblen, T. 1899. *The Theory of the Leisure Class*. 1959 edn. London, Allen & Unwin.

Verṭhaim, A. 1992. *Law and Custom in Hasidism*. Hoboken, NJ, KTAV Publishing House.

Vialles, N. 1994. *Animal to Edible*. Cambridge, Cambridge University Press.

Vogt, E. Z. 1960. On the concepts of structure and process in cultural anthropology. *American Anthropologist* 62(1): 18–33.

Von den Driesch, A. 1976. *A Guide to the Measurement of Animal Bones from Archaeological Sites: As Developed by the Institut für Palaeoanatomie, Domestikationsforschung und Geschichte der Tiermedizin of the University of Munich* (vol. 1). Cambridge, MA, Peabody Museum Press.

Wacher, J. 1971. Yorkshire towns in the fourth century. In Butler, R. M. (ed.), *Soldier and Civilian in Roman Yorkshire*. Leicester, Leicester University Press: 165–77.

Wacher, J. 1974. *The Towns of Roman Britain*. Berkeley, CA, University of California Press.

Wacher, J. 1995. *The Towns of Roman Britain*, 2nd edn. London, Batsford.

Wacher, J., & McWhirr, A. 1982. *Early Roman Occupation at Cirencester*. Cirencester, Cirencester Excavation Committee.

Walker, A., & Long, J. C. 1977. An experimental approach to the morphological characteristics of tool marks. *American Antiquity* 42(4): 605–18.

Walker, P. L. 1978. Butchering and stone tool function. *American Antiquity* 43(4): 710–15.

Wallaert, H. 2012. Apprenticeship and the confirmation of social boundaries. In Wendrich, W. (ed.), *Archaeology and Apprenticeship: Body Knowledge, Identity and Communities of Practice*. Tuscon, University of Arizona Press: 20–42.

Wallaert-Pêtre, H. 2001. Learning how to make the right pots: apprenticeship strategies and material culture, a case study in handmade pottery from Cameroon. *Journal of Anthropological Research* 57(4): 471–93.

Walls, M. 2016. Making as a didactic process: situated cognition and the chaîne opératoire. *Quaternary International* 405: 21–30.

Wang, Y. W., & Pendlebury, J. 2016. The modern abattoir as a machine for killing: the municipal abattoir of the International Shanghai Settlement, 1933. *Architectural Research Quarterly* 20(2): 131–44.

Ward, C. 2008. *An Edge in the Kitchen: The Ultimate Guide to Kitchen Knives – How to Buy Them, Keep Them Razor Sharp, and Use Them like a Pro*. New York, NY, Harper Collins.

Watts, S. 2006. *Meat Matters: Butchers, Politics, and Market Culture in Eighteenth-Century Paris*. Rochester, NY, University of Rochester Press.

Watts, S. 2008. The Grande Boucherie, the right to meat, and the growth of Paris. In Lee, P. Y. (ed.), *Meat, Modernity, and the Rise of the Slaughterhouse*. Durham, NH, University Press of New England, 13–27.

Wayman, M. L. 2000. Archaeometallurgical contributions to a better understanding of the past. *Materials Characterization* 45(4): 259–67.

Weedman, K. J. 2005. Gender and stone tools: an ethnographic study of the Konso and Gamo hideworkers of southern Ethiopia. In Frink, L., & Weedman, K. (eds.), *Gender and Hide Production*. Lanham, MD, Rowman Altamira, 175.

Weedman, K. J. 2006. An ethnoarchaeological study of hafting and stone tool diversity among the Gamo of Ethiopia. *Journal of Archaeological Method and Theory* 13(3): 188–237.

Wendrich, W. 2012. Archaeology and apprenticeship: body knowledge, identity, and communities of practice. In Wendrich, W. (ed.), *Archaeology and Apprenticeship: Body Knowledge, Identity and Communities of Practice*. Tuscon, AZ, University of Arizona Press: 1–20.

White, T. E. 1952. Observations on the butchering technique of some aboriginal peoples: 1. *American Antiquity* 17(4): 337–8.

White, T. 1953. A method of calculating the dietary percentage of various food animals utilized by aboriginal peoples. *American Antiquity* 18(4): 396–8.

White, T. E. 1954. Observations on the butchering technique of some aboriginal peoples: 3, 4, 5, and 6. *American Antiquity* 19(3): 254–64.

White, T. E. 1955. Observations on the butchering technics of some aboriginal peoples: 7, 8, and 9. *American Antiquity* 21(2): 170–8.

White, T. E. 1956. The study of osteological materials in the Plains. *American Antiquity* 21(4): 401–4.

Whiten, A., & Erdal, D. 2012. The human socio-cognitive niche and its evolutionary origins. *Philosophical Transactions of the Royal Society of London B: Biological Sciences* 367(1599): 2119–29.

Wilson, A. 1991. *Practical Meat Inspection*. Oxford, Blackwell Scientific.

Wilson, B. 1978. Methods and results of bone analysis. In Parrington, C. (ed.), *The Excavation of an Iron Age Settlement, Bronze Age Ring Ditch and Roman Feature at Ashville Trading Estate, Abingdon (Oxfordshire) 1974–76*. Oxford, Oxfordshire Archaeological Unit Report 1: 110–26.

Wilson, D. 1986. Excavations at Wortley, near Wotton-under-Edge. *Glevensis* 20: 41–4.

Wilson, D. 1988. Excavations at Wortley, interim report. *Glevensis* 22: 47–57.

Wilson, P. R. 2002a. *Cataractonium: Roman Catterick and Its Hinterland. Excavations and Research, 1958–1997*. York, CBA Research Reports 129.

Wilson, P. R. 2002b. *Cataractonium: Roman Catterick and its hinterland. Excavations and Research, 1958–1997*. York, CBA Research Reports 128.

Witmore, C. L. 2007. Symmetrical archaeology: excerpts of a manifesto. *World Archaeology* 39(4): 546–62.

Wolverton, S., & Lyman, R. L. (eds.). (2012). *Conservation Biology and Applied Zooarchaeology*. Tucson, AZ, University of Arizona Press.

Woods, M., & Woods, M. B. 2000. *Ancient Technology, Ancient Machines*. Minneapolis, MN, Runestone Press.

Woolgar, C. M. 1999. *The Great Household in Late Medieval England*. New Haven, CT, Yale University Press.

Woolgar, C. M. 2006. Meat and dairy products in late medieval England. In Woolgar, C. M., Serjeantson, D., & Waldron, T. (eds.), *Food in Medieval England: Diet and Nutrition*. Oxford, Oxford University Press: 88–101.

Yellen, J. E. 1977. Cultural patterning in faunal remains: evidence from the !Kung bushmen. In Ingersoll, D., Yellen, J. E., & Macdonald, W. (eds.), *Experimental Archaeology*. New York, NY, Columbia University Press: 237–331.

Yener, K. A. 2000. *The Domestication of Metals: The Rise of Complex Metal Industries in Anatolia*. Boston, MA, Brill.

Yeomans, L. 2007. The shifting use of animal carcasses in medieval and post-medieval London. In Pluskowski, A. (ed.), *Breaking and Shaping Beastly Bodies: Animals as Material Culture in the Middle Ages*. Oxford, Oxbow: 98–115.

Yeomans, L. 2008. Historical and zooarchaeological evidence of horn-working in post-medieval London. *Post-Medieval Archaeology* 42(1): 130–43.

INDEX